Dancing with the Tide

"Jimbo" Emmert hand tongs for oysters in a stiff breeze on the South River outside Annapolis.

Dancing with the Tide
Watermen of the Chesapeake

Mick Blackistone

Tidewater Publishers
Centreville, Maryland

Library of Congress Cataloging-in-Publication Data

Blackistone, Mick.
    Dancing with the tide : watermen of the Chesapeake / Mick Blackistone.—1st ed.
        p. cm.
    ISBN 0-87033-532-4
        1. Fisheries—Chesapeake Bay (Md. and Va.)—Anecdotes. 2. Fishers—Chesapeake Bay
(Md. and Va.)—Anecdotes. 3. Chesapeake Bay (Md. and Va.) 4. Blackistone, Mick. I.
Title.

SH221.5.C48 B63 2001
639.2'09163'47—dc21   2001028878

Manufactured in the United States of America
First edition

*To all of those who work the water
and to my wife, Cindy,
and my granddaughters,
Brittany and Taylor Blackistone:
you are my inspiration.*

Man did not weave the web of life,
he is merely a strand in it
What he does to the web,
he does to himself.

—Chief Seattle

Captain Tommy Hallock, a fourth-generation fisherman from Shady Side, Maryland, prepares to empty his pound net of perch, rockfish, sea trout, croaker, flounder, and an occasional bluefish.

Contents

Preface, ix

1 Festivals on the Bay, 3

2 Thanksgiving Changes, 13

3 Fisheries and the Future, 51

4 Bay Fisheries, 65

5 CBF: Controversial Protectionist, 74

6 Watermen Turned Politicians, 82

7 Back on the Water, 103

8 Moving into Spring, 142

9 Bay-Related Legislation, 161

10 Threats to Bay Fisheries, 165

11 Crab Season, 170

12 Hope for the Bay, 189

13 Other Bay Harvests, 209

14 Captain Larry Simns, 243

Epilogue, 257

Index, 263

Preface

Many have compared the Chesapeake's watermen to the cowboys of the wild west—those rugged individuals who spend so much time alone, out in the elements, pursuing their own ideals and not worrying about those of others. While watermen can be attributed those personality traits, in reality, they are far more like our ancestral Native Americans before they were "civilized" by the American government. Like those Native Americans, watermen, by their very nature and profession, are hunter-gatherers—nothing more and nothing less. They do not grow their sustenance, unless we include the few who participate in aquaculture. Rather, they do as they have done for centuries: hunt the Bay and its tributaries for their bounty. If they are successful in their hunt, they will have a plentiful harvest, and if the market price is up, they will be even more fortunate. If they are unsuccessful often enough, they may need to sell their boats and equipment and seek employment opportunities off the water.

Watermen, like those Native Americans, thrive on their ability to read the sign language of their environment. The wind, tide, current, temperature, and time of day and year, along with market price, dictate how they will pursue their "prey." Watermen do not learn from books or lectures how to follow the water or its inhabitants or how to react to their whims. Their ability to be successful often depends on knowledge passed on orally and through hands-on training from generation to generation—or from waterman to waterman—and it works. I am reminded of Robbie Wilson of Tilghman Island, who beat his son C.R. in a boat docking contest more than a decade ago during the festival called Tilghman Day. When I remarked to Robbie, one of the best watermen I know, that then-thirteen-year-old C.R. almost beat him, he nonchalantly responded, "My father and I have taught him everything he knows, but not everything *we* know." Now years later, C.R. and his brother Jason are two of the most respected young watermen on the Bay.

In this regard, it is important to note that there are no vocational schools or classes for young men and women with a desire to work the water. Vocational education programs in our public school systems have done a great deal to establish new generations of technical apprentices in professions from farming to tourism. Through no fault of their own, however, they have neglected the arena of the watermen. There could hardly be a class that teaches adolescents that crabbing is poor when the wind blows from the east, that fewer oysters are caught in a southeaster, that when the eels move the crabs migrate, or that successful oyster diving depends on how well you read the tides. The complex ways of the water and its inhabitants are known only to the few successful watermen, young and old, who were wise enough to absorb information through oral history and experience, filter it for its nutrients, and discard the rest. Despite their contributions to vital attempts at preserving water quality and Bay creatures, scientists can only hold, in my mind, a faint candle to those who relate to the water on a far more intimate level. When asked over a decade ago about what could be done to save the dying oyster fishery and the Chesapeake Bay, renowned oyster expert Dr. Rita Caldwell succinctly responded, "Perhaps we should start listening to the watermen."

I have been in awe of watermen for as long as I can remember. My brother and I were raised on the stories my father passed on about oystermen coming up the Potomac in the early twentieth century, often shanghaiing young men to work the boats up to the Washington markets; or about my grandparents' bar for watermen, located a safe distance from the "big house" on River Springs Farm where the family lived. My father and his brothers were restricted from going into the bar. Nevertheless, on more than one occasion they witnessed my grandfather engaged in combat with an oysterman or three. My father proudly told my brother and me that the watermen left chanting full apologies for their unacceptable behavior.

Children like my father and his siblings, who grew up on riverfront farms in the early days of the twentieth century, often had black "butlers" who watched over them while they played along the shore. These men were there to make sure that the farm boys did not take an unwanted or unscheduled trip to Washington courtesy of the oystermen.

Although my father particularly liked the men of the water and fields, he decided at a young age that he did not have the desire to follow either of those pathways. He was known for his storytelling and he would often take my mother, my brother, and me to Frank Gass's store and bar on Colton Point where he could ply his skill. While my mother played the nickel slot machines and my father talked and traded stories with the locals at the bar, my brother and I would try our best to pull an occasional lucky handle on the slots or wander in and listen to our father, Pinkney Blackistone, at his finest.

While attending the Blessing of the Fleet festival a couple of years ago at the St. Clements Island Maritime Museum on Colton Point, I tracked down Joe Long. Joe and his family are natives who grew up on River Springs Road. I wanted to ask Joe about purchasing a skiff, and he wanted to tell me how, as a boy, he would go to our River Springs farm to listen to my father tell stories about whatever struck his fancy. "I can't tell you the trouble I used to get into for being late to supper because I was over listening to your father. My mother would give me the devil for being over there 'listening to Pinkney Blackistone and his stories.' By God, he was something special to me as a boy." Perhaps growing up with my father's tales, often quite tall, and those of the many watermen I've known, has led me, in some small way, to this book. I have learned that when you are able to read between the lines there is wisdom in those stories.

It is my sincere hope that this personal narrative will inspire its readers to listen to the watermen and to read between the lines. Then, while we all work to preserve the quality of the Chesapeake Bay, we will also work to understand and preserve this important subculture of our society.

This book is really a work in progress—it began in 1989 with *Sunup to Sundown: Watermen of the Chesapeake*. Once again, there were many men and women I did not interview and many cultural disciplines I did not describe. This disappoints me but it would have been nearly impossible to do so.

This work could not have been completed without the help of numerous people and organizations and I am indebted to each and every one: Larry Simns, Betty Duty, others in the Maryland Watermen's Association, *The Baltimore Sun*, *The Capital* (Annapolis), *Waterman's Gazette*, *The Washington Post*, *Bay Journal*, *Bay Weekly* (Deale), *National Fisherman*, the Chesapeake Bay Foundation, and particularly the watermen of Maryland and Virginia. To all those I have not mentioned individually, I am grateful for your contributions to this work and for helping others to learn about the men and women who work the water.

Dancing with the Tide

Competition can become fierce during events like this anchor throw at the Rock Hall Watermen's Festival. Boat docking, crab pulling, oyster shucking, and crab picking are also popular events depending on what time of year a particular festival is held. Courtesy Maryland Watermen's Association.

CHAPTER 1

Festivals on the Bay

Each year thousands of people flock to the shores of the Bay or its tributaries to attend some of the dozens of festivals that, in one way or another, honor the Chesapeake or its inhabitants. From the Rock Hall Watermen's Festival to the Blessing of the Fleet in St. Mary's County, visitors can immerse themselves in Bay literature, crafts, art, photography, music, seafood, and other interests.

For me, the festival that is most relevant to the Maryland watermen is Tilghman Day on Tilghman Island: six hours of island arts, crafts, food, and music, all for the financial benefit of the exhibitors and the Tilghman Volunteer Fire Department. It is a day that gives visitors and residents alike the opportunity to cheer and laugh as watermen compete in such events as an oyster shucking contest, a workboat docking contest, a crab-pot-pulling contest, an anchor-throwing competition, and more. It also marks the time of year when the remaining few skipjack captains join the oyster tongers and divers in final preparation for oyster season on the Chesapeake. While other watermen are still crabbing or fishing, the focus shifts with great interest to the oystermen and the beginning of a new cycle of working the water. In years past, the Bay community celebrated the start of oyster season at Chesapeake Appreciation Days, but state funding shortages and consistently poor weather contributed to a poor turnout.

The state abandoned this traditional event with little remorse and Tilghman Day became the event that brings folks in droves from as far away as Massachusetts to play with Maryland watermen for a day.

On the last Saturday of October I pack up my books for an afternoon of signing as an author exhibitor and drive from my home in Fairhaven to Tilghman. Passing through Annapolis, which competes with Newport, Rhode Island, for the distinction of "Sailing Capital of the World," I cross the Chesapeake Bay Bridge. Out of habit I look over the rails to see if men are working the water south of Sandy Point State Park. I'm not disappointed as I see a crab-potter moving slowly from one buoy to the next. My adrenaline is flowing now and the thought of spending the day on Tilghman leads me to go beyond the speed limit through overly developed Kent Island, across Kent Narrows, south on route 50, and finally through St. Michaels for the last eleven miles to Tilghman Island.

As I drive through St. Michaels I can't help but think of the difference between this upscale enclave—with its historic homes, antique and craft shops, Mercedes Benz and Volvo automobiles, the Chesapeake Bay Maritime Museum, and numerous restaurants—and the towns and communities that surround it, towns like Nevitt and McDaniel, where blue collar locals outnumber white collar weekenders by a huge margin. Passing through these smaller towns, I head toward the end of the peninsula where the new drawbridge joins it to Tilghman Island.

With its collection of watermen's bungalows, the island remained largely unchanged until the 1980s when the state convinced the residents that a new sewage pumping station at Blackwater Point would benefit them and help the Bay at the same time. This new sewer system brought changes to the island that the watermen never expected. They had lived with septic tanks on large lots, among farmland. With the pumping station, the remainder of that farmland would be turned into small lots, overpopulating the island and forever changing the face of their hometown. In the minds of local residents, Tilghman Island today is overdeveloped, overrun with outsiders, too expensive a place for their sons and daughters, and as Captain Robbie Wilson says, "A place where you don't even know anybody's name anymore."

Waterfront towns dot the Chesapeake and an increasing number of people are looking for retirement or weekend homes to buy. In most places, the evolution is a good one, and the newcomers, with sufficient money and a desire to give their new home a facelift, seem to inspire locals to do the same. But in places like Tilghman, Hooper Island, St. George Island, Rock Hall, and other similar communities, the locals are all too often displaced by well-meaning newcomers who are seemingly unaware that they are changing the very environment that was so appealing to them, making it more suburban and destroying the cultural heritage of its inhabitants.

On Tilghman Day, dozens of recreational vehicles fill the fields near Harrison's Chesapeake Restaurant and hundreds of cars with tags from Massachusetts, Connecticut, New Jersey, Delaware, and North Carolina—among others—are parked on the other side of the bridge. As the crowds from these vehicles flock to the fire department and stream down Main Street from the bridge to Dunn's Cove, the islanders aren't giving a thought to the adverse impact. It's ironic that while the locals need the money engendered by this event for their impressive volunteer fire department, they also tend to resent the fact that so much of the support comes from those who now use the island as a weekend or vacation home.

As I set up a table to sign my books in a shared space with Island Treasures, a gift shop owned by Leslie Harrison, wife of Captain "Little" Bud Harrison and daughter-in-law of well-known Buddy Harrison, I wonder how outnumbered the watermen are by the throngs of Tilghman Day enthusiasts who have come to celebrate with them. I fantasize that the outsiders are like visitors to a circus or like the Romans flocking to the Colosseum; they gather for one day to watch the watermen compete against each other and against Mother Nature and then disperse at the end of the day, leaving their plastic beer cups from one end of the island to the other as evidence of their presence. While it is a day of fun and competition, it is also a day when I can look in the eyes of Robbie Wilson, Bill and Jean Cummings, and even Bud Harrison and see that they feel they've lost hold of their world.

Volunteers prepare seafood for the Watermen's Association food tent at the Maryland State Fair, an exhausting ten-day event that requires a lot of participation and coordination. It's worth the sacrifice because this event will draw tens of thousands of attendees who will provide a strong source of association revenue. Courtesy Maryland Watermen's Association.

In my own neighborhood off Herring Bay in Fairhaven, oysters provide an excellent excuse to get out and socialize. Everywhere you turn in late October, there is a sign or flyer advertising an oyster roast or a ham and oyster dinner. Churches, civic associations, and fire departments serve up what is a delicacy to many and an abomination to others. This cornerstone of the local economy is served fried, frittered, puffed, stewed, steamed, roasted, or raw. Purists prefer raw, freshly shucked oysters. "Bite 'em once to kill 'em, and let 'em slide," the saying goes.

Perhaps the biggest oyster event in my neck of the woods has been at the Deale Fire Hall. For the past thirty-eight years it has been an annual tradition, although it actually started some fifty years ago down the road at Albin Parks's Tavern, now the popular Skipper's Pier. On the first Saturday in November, over three hundred oyster fans show up for local community festivities.

It would be unfair to ignore what was once the primary event on the Bay for celebrating the start of the oyster season: Chesapeake Appreciation Days, held for the last time in the midnineties. CA Days, as it was familiarly and affectionately called by thousands of the region's residents, was a weekend of food, music, craft displays, how-to demonstrations from oyster shucking to diving, skipjack rides, educational discussions and storytelling, and, most importantly, skipjack races on the Bay starting from Sandy Point State Park and running under the Chesapeake Bay Bridge. It was a captivating sight to see these legendary vessels, under full sail, competing for the prized trophy of fastest boat in the fleet—at least for a day.

Skipjacks used to race with pride during Chesapeake Appreciation Days before the well-known event was canceled by the state due to a lack of funds. Courtesy Maryland Watermen's Association.

Russell Dize, captain of *Kathryn*, is a widely respected fourth-generation waterman, a businessman, a family man, an original founder of the Maryland Watermen's Association (MWA), and a Methodist from Tilghman Island. Like so many of the captains, he did not want his son to follow the water. He and his wife Joan wanted their son Rusty to get past the instant gratification of quick money—by way of one good harvest—that so many young men thrive on. Russell calls it "temporary money" that only leads most of them to a difficult life of trying to make ends meet. Rusty was never allowed on Russell's boat except during the skipjack races at CA Days.

Russell's sister Deenie made her home a welcome port to me during the writing of *Sunup to Sundown*. Her sons followed their late father Gene in looking for quick money and "the end of studies." "I know Rusty probably suffered a little bit because following the water is what so many boys do here on the island, but Russell was right. Even though it was hard to keep him away, he's better off for it," Deenie said.

CA Days, like so many other festivals, fell to bureaucratic and political mediocrity; turf battles within the managing organization, Chesapeake Appreciation Days, Inc.; and economic hard times. The Maryland Department of Natural Resources had funded a sizable portion of the event and when it pulled out for budget reasons the Maryland Watermen's Association took over the task of management for the corporation. The MWA was successful for several years, but repeated inclement weather and the lack of state funding finally led to the festival's demise. Many were saddened, frustrated, or furious with the announcement, but emotions did nothing to inspire resurrection of the event.

What did come out of CA Days was a new, independent organization, the Oyster Recovery Partnership (ORP), developed by the corporation under the tireless leadership of former Executive Director Bob Pfeiffer. The ORP is a nonprofit organization with one mission: restoration of the Bay's oyster resources. The organization is made up of watermen, seafood processors, scientists, environmentalists, private industry, government, and members of the general public. Working with the University of Maryland Oyster Hatchery at Horn Point in Cambridge, Maryland, the ORP grows its own oyster spat (young) for oyster bar restoration. The goal is to create critical mass oyster plantings throughout the Bay to supplement natural oyster spawn. The ORP planted forty million oysters in various locations in the year 2000 and a similar number is planned for 2001.

Many of the other oyster seed programs that supplement the ORP's efforts have resulted from the creation of the Maryland Oyster Roundtable, which convened initially in 1993. This group's efforts have been bolstered by the United States Army Corps of Engineers which authorized the allowance of a maximum of $7.5 million in federal project funds. Administered by the Baltimore district of the corps, the proposed seven-year projected activities include

- creation of new oyster bars and rehabilitation of existing nonproductive bars,
- construction of seed bars for production and collection of seed oysters or spat, and
- planting of hatchery-produced spat and spat harvested from the seed bars onto new bars.

These activities were put in place in 1997 and concluded in 2000, with monitoring of the implemented projects continuing until 2003. Estimated cost of the corps-related oyster recovery project is $3.3 million, with 25 percent coming from the State of Maryland.

The members of the oyster roundtable also received input from the state's Chesapeake Bay Program, when Governor Parris Glendening announced that the program's goal was to increase the oyster biomass "tenfold by 2010," an ambitious aim by any standard considering the state of the fishery.

And while residents of the Bay wring their hands over the cautious optimism for recovery of the oyster population, watermen—the eventual victims or beneficiaries of so many efforts—continue to live as they always have, from harvest to harvest.

The men and women who work on the water are known for their passion for competition, and festivals provide a forum. For over twenty years oyster shuckers from all over the United States have converged on St. Mary's County, Maryland, for the annual Oyster Festival where the U.S. champion is crowned; the winner goes on to the World Oyster Opening Championships in Galway, Ireland. On October 17, 1999, George Hastings of Severn, Maryland, walked away with the St. Mary's championship. Hastings defeated eleven male competitors and the women's champion, Clementine Macon of Urbanna, Virginia, to win the competition. Macon was actually seven seconds faster than Hastings, 2:12 to 2:19, in shucking two dozen oysters, but she lost because of penalty points assessed after the judges reviewed for cuts and nicks on the shucked oysters. Unfortunately, in a surprise move, Guinness, the Irish-based ale maker and sponsor of the international championship, decided to hold its own shucking events that will send an American other than St. Mary's champion Hastings to Ireland. Although Guinness has control of the Irish championship due to its financial sponsorship, St. Mary's representatives say they have a "patent, a trademark, on the words 'national championship.'" The discussions continue as watermen and country folk from southern Maryland watch helplessly for the outcome of the debate and wait to discover who is the world's fastest oyster shucker, or, as they say in Ireland, oyster opener.

The original inspiration for many of these festivals came from the Maryland Watermen's Association, which represents the hardy souls working the water, as well as their families, on issues ranging from legislation and regulation to insurance. The MWA, aside from being an inspiration, is also the primary vehicle behind waterman-related events and exhibits for the general public. Since its inception in 1973, the association has been run by Larry Simns, president, and Betty Duty, association administrator, assisted by an ever-changing staff of loyal employees who manage to carry a heavy load of work during the time they serve the organization. To ensure all officers are working watermen, the association has had elections of officers on a regular schedule over the past twenty-seven years. The members always reelect Simns, a gillnetter and charter captain, as he constantly demonstrates his political savvy, negotiating skills, and leadership before Congress, the state legislature, regional and state fisheries councils, environmental groups, and the general public. Besides, no one else is willing to devote time to the job.

I have considered Larry and Betty good friends for over twenty years, and while I represent recreational boating interests in my day-to-day professional employment, I occasionally go to Larry Simns for an objective, insightful perspective on a potential problem or issue. A few years ago, both the commercial watermen and the Recreational Fishing Alliance (RFA), usually competitive opponents, had a common interest in "removing" a member from the Atlantic States Marine Fisheries Council; the person had been appointed by New Jersey's then-Governor Christie Whitman. It was my dubious distinction to receive timely but coincidental telephone calls from Jimmy Donofrio, executive director of the RFA, and Larry Simns about helping to secure the "removal" of the member through the governor's appointment

Betty Duty has been in charge of the Maryland Watermen's Association office and fund-raising events, and has been Larry Simns's right hand for over two decades. She is widely respected by those in the public and private sector as a "get it done" type of person, and many feel the MWA would be lost without her. Courtesy Golden Memories Studio, Inc.; copyright © 1999.

process. Two adversaries seeking a mutual goal can often agree whether they realize it or not. New appointments were eventually made, and the unwanted member was off the council. To this day, I'm not sure either Jim or Larry knew they were both working for the same goal.

It is a rare occasion when Larry Simns seeks assistance on MWA issues. He moves with subtle, soft-spoken authority, confidence, and energy on behalf of his colleagues and is usually victorious, regardless of the arena. He and Betty move as well in environmental and recreational boating circles as they do among the watermen. I have admired and respected them for many years for their skills and accomplishments. As I observe their work schedules, persistence, and tireless dedication, I wonder what will become of the Maryland Watermen's Association as these two treasures of the Chesapeake move closer to retirement. There are, of course, younger men and women following in their footsteps—Kenny Keen, president of the Calvert County Watermen's Association comes to mind—but leading the MWA is usually a thankless job that requires its president to spend time in the office as well as on the water, a juggling act of immense proportions with few rewards.

CHAPTER 2

Thanksgiving Changes

For many Maryland and Virginia residents, Thanksgiving means change. We finish raking leaves as trees shed to skeletons against a full moon. We have started holiday shopping, and we have even had a fire or two to take the chill out of houses draped in mist under cloudy skies. Wading in the Bay ended weeks ago, despite being blessed with an Indian summer, and many boatyards are looking for their diehard customers to finally winterize and put their boats into winter storage. Sailors still dot the Bay and rivers, and the frostbite races attract enthusiastic participants as cooler air fills their sails and moves them beyond frustrations lingering from the doldrums of August.

For watermen like Irving Maddox, Kenny Watts, Jackie Russell, Bob Evans, Kenny Keen, and Robbie, C.R., and Jason Wilson, Thanksgiving and November 30 mark the end of crabbing season and the transition into clamming, oystering, and fishing. According to state regulation, all crab pots must be pulled by the last day of November, and while some men have already started working other harvests, a few men like Kenny Watts continue to crab until Thanksgiving. The Wilson men pulled hundreds of pots the week before the holiday.

PATENT TONGING WITH KENNY KEEN

On the Friday morning before Thanksgiving, it is still dark as Deale, Maryland, sleeps, and I enter the vacant parking lot at Calypso Bay to meet Kenny Keen. As I park my car, I am reminded once again why I titled my first book on the watermen, written a decade ago, *Sunup to Sundown: Watermen of the Chesapeake* and that I promised my wife, Cindy, that I would be home by sundown. I look at my duffel bag containing lunch, foul weather gear, my camera, a notebook, and pencils. It is thirty-two degrees, lower with the windchill. I learned the hard way years ago that ballpoint pens freeze up and a pencil is by far a better instrument under these conditions. As I spot the only vehicle in the parking lot and pull up to the white pickup, Kenny rolls down the window and asks me how my trip to Chicago was the day before. After a few minutes of small talk he tells me to follow him down the road to Pier 44 on Rockhold Creek. Concerned that I still look half asleep as I get out of my Jeep, he suggests that I go down the road to the 7-Eleven to get a cup of coffee. Good idea, I think, but I don't want to impose on his time. I learned a long time ago that to watermen time is money, and they both are of equal value.

"I'm okay," is my first response to his suggestion. I realize that he is being gracious but I'm reluctant to overstep my bounds.

"No, go ahead. You'll be back in five minutes," he insists.

"Okay. Can I get you some?" I offer without hesitation.

"No, thanks. I've already had my twenty ounces. I won't blink for about three hours!"

When I return, I lock the Jeep and begin to walk to the boat. Kenny instructs me to follow him. It's dark and there is a thin sheen of frost on the pier.

Keen, thirty-eight, is president of the Calvert County Watermen's Association. He crabs and oysters from his boat, *Long Shot,* a forty-five-foot fiberglass baybuilt, a boat designed with a high bow for slicing through the water and a low stern to make it easy to work off the stern or amidships. The week before, she had been stripped of her winder, an automatic winch that is used to pull crab pots, and of the overhead canopy used to block the summer sun and often to store pots. Now she is retro-fitted with a mast, pulleys, lines, hydraulic hoses, culling board, and pat-ent tongs for oystering. Patent tong rigs were first introduced in 1887 when Charles L. Marsh, a Solomons, Maryland, blacksmith, made known his first design of the deepwater tongs. Over one hundred years later, tongs all look basically the same, although watermen from differ-ent parts of the Bay may add their own unique rigging coming off the main mast.

As I board the boat and enter the cabin to store my gear, Keen has the heater running and the VHF radio on. I immediately recognize the voice of Tilghman waterman Robbie Wilson. Robbie is conversing with another waterman as both men prepare to haul their crab pots. Novem-ber 30 draws near. Robbie asks the other fellow, "What do you want to do when you pull these pots up? Go fishin' or go to the ocean to sun-bathe?" He shares jabs back and forth with other men who respond to the questions, including sons C.R. and Jason, both ending their season of crabbing as well. Ironically, without a hint of knowledge, although with a good suspicion, Robbie gambles on Kenny's listening to the Tilghman banter:

"Did you all read that letter from Kenny Keen's wife in the *Water-men's Gazette?*" he asks, referring to the following letter that Karen Keen wrote in February 1997 to the editor of the Maryland Watermen's As-sociation's publication in response to a request for submissions from the wives' perspectives:

WATERMEN'S WIFE GROWS
TO APPRECIATE NEW LIFESTYLE

Though born and raised on the Chesapeake Bay, the Waterman was a person I was not familiar with. He seemed to be just a guy on a boat who rode past our house in the early morning. Of course I was always asleep so I never got a glimpse of him. Four doors down lived our neighbor Mac McNasby, who owned McNasbys Oyster Company in Annapolis. Little did I know my future would cruise in that direction.

When I met my husband Kenny he was not working on the water, although he had in the past. We fell in love and got married to live happily ever after. Soon after we were married, Kenny came home from work—he was working for my dad—and announced he was going back to his real love: working as a Waterman. I looked at him and thought to myself, "You've got to be kidding!" He wasn't. He borrowed some money to buy his first boat.

In all his excitement, the day finally came when he took me to see his new vessel. I took one look and scratched my head, thinking that there was no way it would ever stay afloat. But lo and behold, he fixed the boat up, repaired the engine, and added a roof so the crabs wouldn't bake in the sun.

Meanwhile, our front lawn had turned into a crab pot assembly line. What had once been my view of green grass and flowers was now galvanized wire, zincs, floats, cord and cull rings. Ken worked for hours creating crab pots, and I would watch and wonder how this was possibly going to work out.

The time came to set the pots, so Ken bought the bait and let the pots sit for a couple days, which meant he was home again. By this time I was eager for him to go to work, but at least I could see my front yard again since the crab pots were gone.

Then the bad news. A red tide came in and killed the bait. Ever the optimist, Ken said, "No problem! I'll just move the pots and re-bait." Needless to say, I was feeling a little shaky about this whole Waterman career thing. Finally, the first good day of crabbing came and I thought, "Hallelujah! Maybe, just maybe, this will work!"

Now after ten years of marriage, my education of the Waterman's way of life has blossomed. What I thought would never get off the ground has proven me wrong. Hard work, long hours, patience and a little help from Mother Nature has made this a special way of life.

My husband goes to work in a good mood and comes home in a good mood—what else could I ask for?

—Reprinted courtesy Maryland Watermen's Association

"It damn near brought tears to my eyes," Robbie offers over the VHF airwaves of channel 6, the "watermen's channel." Kenny laughs at the remark as *Long Shot* moves out of her slip and into Herring Bay. I know from past experience that he will laugh a lot during the day—he always does—and I think his voice is hoarse because he laughs so much, so often, especially in the company of other watermen.

Moving through the darkness of the Bay, I gaze out the starboard side at Fairhaven to see if I might see a light in any of the houses, particularly mine. Predictably, the small Bay community of former summer residences rests in darkness. Kenny comments that several watermen live in Fairhaven, and I confirm that Steve and Scottie Smith and Dave Watts live there and that Kenny Watts lives right up the street from me. Keen lives about ten minutes south along the eighteenth hole of the Twin Shields Golf Course, which is owned and operated by his wife, Karen Shields Keen.

Kenny tells me we're going up to Annapolis to Hacketts Bar, an oyster bar located off the David Taylor Naval Station. The radio remains on and represents the only means of communication these men will have with one another or with land all day. It is the source for neighborly conversation, inquiries about how to best fix a problem, help for a tow or injury, discussion of the day's issues, and more. Kenny now holds the hand mike and joins the rotating conversations, some with the Tilghman watermen and others with men from Chesapeake Beach.

Keen takes a break from the radio to tell me that Martin Legg of Rock Hall, Maryland, built his new culling board. He's excited about it but will soon find that culling the oysters that are dropped on the board from the tongs will scratch the shiny aluminum. When they are culling or sorting oysters watermen select oysters of legal three-inch size, and they push the rest—the trash—overboard. The scratches won't bother Keen, but they will make it more difficult to push the trash overboard. To solve the problem he will put a sheet of stainless steel over the top of the board. Culling boards are frequently made of iron, but Keen thought that he would try aluminum for his replacement. The soft metal barely held up three hours on the first day before deep gashes appeared on its surface.

As we head to Hacketts Bar he says, "It's a big bar and there will probably be a mess of boys up there. I'm not too worried about doing much on my first day. I just need to get the kinks out and make sure

A patent tong fleet off Annapolis working an oyster bar.

everything works right." He will emphasize several times during the day that the most important job he has right now is to make sure the boat and rigging are right and that the tongs and hydraulics are working properly.

"When you're out here by yourself it's a much nicer day if nothin' goes wrong. A couple of years ago around Christmas I got a line wrapped around the prop and had to strip down to my underwear, go overboard, cut it off, and get the hell out of that water. Christ, it was cold. Then while I'm standing in the cabin thawin' out, a woman patrol officer comes by to give me a ticket for working past the 3:00 P.M. closure time. Well, I had to get dressed before I could tell her what happened and she gave me a ticket anyway. I had to take off a day of work to appeal the ticket in Queen Anne's County. The judge let me off and told me to stay dry. I still see her on the water and we laugh about it now," he says.

I ask Keen if working on the Bay is the only work he's ever done. "I've been on and off the water for eighteen years. My grandfather was a gillnetter up in New Jersey, and I learned a lot from him. I loved gillnettin' but I didn't feel like chasin' fish all day. I like oysterin'. There's no bait, I can work alone, and these oysters don't swim! But I'm like most guys, learn from the older or better watermen, employ a lot of trial and error. Truth is, you've got to want to do this job. I've done other things between times, even took time off to work on the golf course. I've never been so unhappy in my life. I went into a little depression, started drinkin' and partying all the time with guys who hung around the course to escape. Karen and I talked and we both knew I had to get back on the water. I help out in the winter but it's better this way. She does her thing and I can do mine."

We are moving with a strong southerly breeze in three-foot seas up the Bay. It's rough and Kenny complains that he should have stayed up in Rock Hall and worked there for a few days. He can oyster from sunup until 3:00 P.M. with a fifteen-bushel limit. If I had a license, I could count as crew and he could bring in a maximum of thirty bushels, still by 3:00 P.M. "Man, I hate this wind from the south. Makes for a rough-as-hell trip, and for some reason we don't catch many oysters on a day like this," he explains.

On the radio we overhear Robbie Wilson talking about a Sea Tow boat that ran through his stationary pound (fish) net. I could tell that men from all over the Bay had been talking about the incident off and on all morning. Robbie is upset. "He's flyin' home but he better sober up if he can't see my net when he runs this water all the time," he states to whoever is listening. After a minute or so of venting, he calmly states that the man ought to offer to pay for the damage and he assumes he will. End of conversation.

Robbie and his sons, C.R. and Jason, switch off the open channel. "They do that all the time. They break off channel 6 and by some code or something end up on another channel to tell their secrets about how they're doin' or what they're plannin'. I'm gonna find that channel one of these days!" Keen says with a laugh. "When we're crabbin' in the summer and one of them says, 'Go to that other channel,' everyone will stop what they're doin', run to their radios, and try to find 'em. Damn if you can though," he laughs again.

The MWA is fortunate to have younger watermen coming along who enjoy getting off the water and working for the good of the industry by attending a myriad of meetings that impact the way watermen do business and the way they live. Keen says that he only got involved a little over a year ago and that he only got to know Larry Simns and Betty Duty during that time. Regardless of his lack of involvement in association duties prior to 1998, he has worked hard representing the state's watermen and Simns on the Oyster Recovery Partnership, the Bi-State Blue Crab Advisory Committee, the Oyster Roundtable Steering Committee, the Patuxent River Commission, and more.

In April 1998 he represented the Maryland Watermen's Association at a two-day conference I put together with Richard Gutting of the National Fisheries Institute (NFI), which represents commercial fishing interests on a national level. While many of NFI's members are seafood processors, large factory trawlers, and restaurants, others are independent watermen. As vice president of government relations for the National Marine Manufacturers Association (NMMA), I represent the recreational boating industry trade association whose members manufacture over 80 percent of the boats, engines, and related accessories made in the United States. Gutting and I thought that if we could agree to disagree on certain issues, we could jointly host a conference on the environment and fisheries.

A select group of leading scientists, representatives of the National Marine Fisheries Service, and an equal number of participants from the recreational and commercial industries were invited. The group addressed commercial gear, recreational and commercial fishing quotas, and the impact of both on fish habitat and species conservation and recovery efforts. A plan of action was developed that both industries agreed to present to their members. In addition, they made a pledge to continue individual partnership efforts to improve both environmental quality and in particular to boost populations of large pelagic species like tuna, shark, and swordfish. Keen represented the independent watermen very well; it was at this meeting that I first met him and saw him in action.

Keen and I reminisce about such activities and issues, including the NMMA's efforts to encourage the Marina Operators Association of America (MOAA) and the Marine Trades Association of Maryland

(MTAM), which represent recreational boating businesses, to help local watermen identify boat slips for their use. This is a serious problem because the watermen cannot compete with large recreational vessels whose owners will pay up to $3,000 annually for a boat slip.

The sun is a red ball of fire showing itself over the edge of the horizon as we approach Annapolis and Hacketts Bar. Sunup means that the tongers from Tilghman Island already on the bar can begin work, and they do so without hesitation. Keen has turned on his fish finder and shows me the outline of the oyster bed. "She's plenty big, and we'll see what we do today as I work the bugs out of the rig. I'm not counting on much. I just hope everything works right and I can begin to get the feel of tonging. It will take a while and I wanted to get going even though I could have crabbed until November 30—that's when the pots have to be out of the water," he says.

It's cold and the wind is blowing fifteen to twenty knots from the south over three-foot seas. In the pitch and roll Keen notices that the heater stops working.

"Damn it. I've got to get this fixed for you," he says, staring at the grate on the side of the wheelhouse.

"Don't worry about me," I reply sheepishly. "I'll be in and out all day anyway."

He removes four screws from a cover below the steering column to reveal a jumble of wires in a variety of colors. "No, I've got to fix it anyway. In another month that wind and cold will be blowing right up my ass, so I've got to get her straight now," Keen explains as he sticks his head in the opening and searches for the wires that lead to the heater switch.

I remind myself that it's best to keep quiet and out of the way when something happens on a boat. These men are used to working independently and if they need help, which doesn't seem to be very often, they ask for it. Within three minutes Keen is screwing the cover back on and instructing me to flick the second switch down if I want the heater on. He looks frustrated that this kept him from what he came to do. I remain quiet and he leaves the cabin, with me in his shadow, to turn on the engine for the hydraulic rig.

Hoses carrying hydraulic fluid to work the tongs are draped off the mast and rigging and surrounded by winches and pulleys. All of this leads to a pair of patent tongs which look like giant rakes about three feet long with over a hundred claws on the ends to grab the oysters from the bottom when the tongs are lowered into the water. Keen moves to stand behind the three-foot by five-foot culling board which is bolted to the floor midway between the cabin and the stern and offset to the starboard side. From this position he and other patent tongers work as maestros orchestrating a day of oystering. With the culling board in front of him waist high, the boat's throttle and tiller twelve inches from his right hip, and two foot pedals—one to raise and lower the tongs and the other to open and close them—Keen is ready to make his first run on the oyster bed. "I'm a little rusty but I'm excited. I love doing this," he states with the enthusiasm of a ten-year-old boy with a new toy.

I, too, am excited. It has been ten years since I was last out on a working vessel and this is my first time with Keen. We both watch as he lifts the rig from the culling board and shoves it out over the starboard side. Using the foot pedals, he lets it crash open fisted into the water. Seconds later it resurfaces with a handful of oysters. The tongs stop a foot off the side of the boat and about a foot above the surface of the culling board. They have already been adjusted to be within easy reach of the captain but unable to swing in far enough to hit him as he pulls the tongs over the culling board and opens them with the other pedal. The first oysters of the year crash against the new aluminum surface. As he raises the tongs and shoves them out over the side to send them below again, he flashes me a big grin and laughs. The process will continue until we take up to fifteen bushels of oysters or reach 3:00 P.M. Kenny's feet move like Elton John's across the foot pedals; his right hand adjusts the throttle and tiller to keep us drifting in the right direction and then reaches up to the patent tongs to pull them over the culling board or to send them back over the side.

As soon as the oysters hit the board and the tongs are sent searching again, Keen moves simultaneously with both hands through the modest pile of perhaps three bushels of oysters and shells that lie be-

Kenny Keen raises his patent tong using a foot pedal and guides it over the culling board, releases his catch, and sends the tongs back overboard before culling the oysters.

fore him. As he spots an oyster obviously over the three-inch legal length he tosses it without looking into a bushel basket sitting on the engine cover to his immediate left. Everything he needs to carry on his trade is within an arm's length. When rummaging reveals a smaller oyster of questionable legal size he reaches across the pile to an apparatus he uses for measuring. It is a customized handle with a welded letter F attached to it. The opening in the letter is three inches, the minimum size for an oyster to be taken legally from the Bay. He quickly passes the oyster through the opening and it does so without striking metal. He throws it overboard. "If it doesn't hit the two bars it's too small. You can get a nice ticket for taking illegal oysters, so I play by the rules pretty close. Oh, one or two will get by me but it's usually not too intentional. Fact is, it's usually about fifteen minutes before quittin' time when I want to get the hell off this bar," he says and I think the explanation makes sense.

Two men from Tilghman are working a double patent tong rig about forty feet from us. In this case, each man has his own set of controls for the tongs and the captain also works the steering and throttle of the boat. Keen moves *Long Shot* away from these men as two more boats arrive. Within thirty minutes of our arrival on the oyster bar, ten boats have joined us, all moving slowly in wide circles like sharks preparing for a feeding. The names of the boats around us reveal something personal about each of the captains. *Bad Investment, Lady Jordan, Cody Thomas, Miss Marley II, Leslie Ann* from Tilghman, and *Henchman* from Chester, Maryland, are among the vessels scavenging the Bay bottom in rolling seas.

Keen culls the oysters on the board and remarks, "Lots of little ones. Don't do me any good today but it's an encouraging sign for tomorrow." As he's finishing his statement I look up from the culling board to notice that *Miss Mornin'* has drifted to within eight feet of our starboard side, heading straight for Keen's transom. I yell to Kenny who looks over his

Kenny Keen culls oysters aboard *Long Shot*.

shoulder and throws the throttle forward to get out of the way. "Damn it. I didn't see him and he didn't see me. We would have been in a hell of a mess. You've got to keep the boats moving or you pull up the same oyster shells, but with this wind and your eyes on the culling board it's hard to look out for the other guys." He's quickly back to work.

As I join him in culling oysters, dodging the tongs while I work, he surmises that the Natural Resources police may be out today checking on the catches. "A lot of the guys are tough on the police. I try to be nice so they might cut me a little slack. Sure don't want a ticket because a smaller oyster got by me."

I ask him how much a ticket would be and he admits he doesn't know because he's never gotten one. "It's easier and better to follow the rules. Some guys think getting a ticket is part of doing business. I don't. That's bull when you consider a fine can cost you damn near a day's pay and you earn a bad reputation with the police. Then you catch it all the time from 'em. Nope, it's not worth it to me."

When I went clamming with Larry Simns years ago, he illustrated how watermen identify good clam beds by landmarks such as a tree, a house, a tower—signs known only to them. Now, in the midst of a dozen patent tongers working one bed off Annapolis, I watch Kenny Keen toss a makeshift buoy, an empty Texaco oil bottle tethered to a weighted line, to mark this area as his territory. "It's an informal and unofficial marking system which means I like this area and intend to work it for a while. Sometimes a guy will come in and work it for a few minutes until I chase him off, and he'll do the same to me. It's a code of the west kind of thing. Anyway, by watching the buoy I can get an idea of how we're moving over the bar," he states.

The wind continues to pick up from the south, so Kenny drops an automobile crankshaft tied to a line off the stern of the boat. "With the tide going one way and the wind another, this will help her stay steady," he explains and heads back to the culling board, his foot pedals, and the tongs. I spot an antifreeze bottle—someone else's territory—off to our port side and Kenny states matter-of-factly, "I'm just working the bar back and forth getting to know it. There are some good oysters here—that's the reason for the buoys. These bars change every year, so you've got to work them to become familiar again."

Keen explains that there are strict limits in all the fisheries regarding catch, gear, and hours of operation. For a patent tong rig like *Long Shot* the limit is 15 bushels per licensed man. Oyster divers can catch up to 30 bushels with two men diving; hand tongers can catch up to 30 bushels with two men working the tongs; and skipjacks can take up to 150 bushels.

"I know a diver who worked eight years and never caught his limit, then one day he had a different view of the tide and he's had good luck ever since. When you're diving for oysters you really have to dance with the tide. It's not the captain on a divin' boat that leads to success; it's the diver. Hell of a job."

As I help cull oysters in an attempt to earn the information Keen is passing on to me, *Henchman* is working next to us. "He's a good boy, there," Kenny says. "He was going to rig a conveyor belt to his cullin' board so the shells would walk right off the end. It was a hell of an idea. You start thinkin' of all kinds of things because you get so damn tired of shoveling shells."

To get our minds off the monotony of the culling routine, I mention the concerns recreational boating and environmental groups have with the longline fishing industry, particularly with the factory boats. We both share our concerns over the prominent bycatch issue (when watermen unintentionally kill one species of fish while harvesting another) and that quite often the problem with fishing is not allocation/quotas and overharvesting but poor water quality leading to poor spawning, feeding stocks, and fewer adult fish, with a lot of fishermen racing for them.

"I wish they would stop blaming the decline in fisheries on fishermen and overharvesting and devote more time to good science, research, and pollution controls," Keen says. "Spawning grounds are gone and juvenile fish aren't surviving because of pollution—that's the impact on viability of the species."

I agree but give him my opinion that perception and public opinion will continue to point a finger at gear types like longlining, regardless of the combination of factors that impact mortality. Pictures presented by environmental groups concerned for turtles, dolphins, sharks, and other bycatch victims are what's hurting the industry. He reluctantly agrees saying, "We don't do a very good job of promoting the good things we do in this industry."

At eleven thirty Kenny loses the hydraulic steering by the culling board. We say "damn" in unison as he quickly moves to the steering station behind the cabin, in front of the culling board and mast. He steers the boat away from other tongers and looks underneath the steering apparatus to see if he can switch valves and solve his problem. That wasn't the problem. A small pin that connects the steering rod to the hydraulic valve had fallen out. Transferring pins takes five minutes and we're back to tonging.

There are more buoys now indicating that the men are becoming familiar with the bar today and claiming the better spots to work. Coke bottles, oil bottles, antifreeze bottles all serve the same purpose.

At eleven forty-five a Tilghman man has his fifteen-bushel limit and another Tilghman boat with a double rig has its thirty-bushel limit. They are heading south toward the Eastern Shore and home. We have six bushels and Keen laughs saying, "We'll be the last boat to leave this party today."

He asks what kind of music I like, and when I admit to liking just about anything, he suggests a couple of stations we can turn on to have "a little diversion." He also asks politely if I'm hungry and reminds me of the pastrami and cheese sandwich he bought me from the 7-Eleven. "Man, I love those things. I'm hooked on 7-Eleven sandwiches. Keep your tuna fish and try one," he insists. I agree and wait for him to leave the culling board for his sandwich, soda, and a little music. After fifteen or twenty minutes I realize that he's no different from any of the other men I've been out with. He's not eating until we've reached the 3:00 P.M. deadline or fifteen bushels. He's not thinking about music, and he's fine if I decide to do either. I don't.

By two o'clock we're the only boat on the bar. We have ten bushels and an hour to go. "Well, they say a good run is better than a last stand.

Just ask Custer!" he yells and laughs. "Hell, we're doin' great for the first day. Better than I hoped."

At three o'clock he pulls the crankshaft up from the water behind the transom and rests the tongs on the culling board. Moving toward the cabin, he stops to count thirteen bushels and smiles. Once inside, he takes off his foul weather gear and relaxes at the wheel, grabs a banana, and turns the boat south toward Herring Bay and home.

"I'm gonna open her up a bit to get out of here, but it's gonna be a rough one. Let's see how she handles this weather." Ten minutes later we're running head-on into four-foot seas, and I'm holding on for dear life. "Man, I hate this wind. I swear I'm getting a turbo diesel and getting rid of this crap. It will take us forever to get home and we're gonna beat our brains out all the way to Deale."

Driving into a headwind and choppy seas is an unpleasant experience even on a forty-five-foot boat built to take the Chesapeake with ease. Everything is flying around the cabin as the boat surfs down one wave only to ride up the next one and pound into the trough with the sound of a sledgehammer. It would take us more than an hour in this weather to get into Herring Bay and Deale harbor. As we move along, Kenny cusses as one wave after another crashes over the cabin showing us leaks along the windowpanes he'd never noticed before.

"I guarantee the weatherman is reporting a nice sunny day with a slight breeze from the south on the television. He ought to be out here on this roller coaster," he says as we are dropped over five feet off the crest of a wave.

Once in Deale harbor, we head for Pier 44 Marina where Kenny's keeping the boat for the time being. As he backs the boat into the empty travel lift slip, I gather the lines in an effort to make the fastest exit possible. When the boat is tied, he gets his truck, and we load up thirteen bushels that will now go to his buyer on Broomes Island, about forty minutes south of Deale. He's promised half a bushel of the thirteen to a private customer. These oysters will be modestly skimmed off each bushel basket with no real harm done to the quantity promised to the buyer.

At four thirty I'm driving home realizing that Kenny Keen will be home about six o'clock—a twelve-hour day for $260 in oysters, several minor panics over loss of steering and loss of heat, and seas that pounded him until the three o'clock deadline. Yet he was tickled with his first day of tonging, and I was too, though I never did get to hear that country radio station he suggested hours before.

HAND TONGING WITH BOB EVANS

Hand tonging for oysters is an altogether different experience from working a patent tong rig. Hand tongs have slightly smaller tongs; they're shaped in the same clawlike fashion as the patent tongs, and they are mounted on long wooden shafts, the length depending on the depth of the water over the oyster bed. A bolt connects the shafts, allowing them to work like scissors. Standing on the boat's washboard holding the shafts, the hand tonger lowers the tongs into the water, touches the bed with open tongs, and pulls the shafts apart, bringing the tongs together before lifting them up to the boat for the oysters to be culled. The process is the same as that of the patent tong rigs except that in this case it's done with manpower rather than hydraulics.

One of the boys on a skipjack told me several years ago that you can tell a hand tonger because "he's got big arms and a big chest, but a little tiny head." This lighthearted comment made in jest, like many others these men make about each other, is a lot of fiction. In fact, hand tonging takes strength, finesse, balance, and tremendous expertise, not to mention knowledge to work the tides, currents, breeze, and beds.

Bob Evans of Churchton is a barrel of a man and one of the smartest, friendliest, and most generous men I know. He is quick to volunteer, whether the task is to shuck oysters at a dinner gathering or to represent the Maryland Watermen's Association at the Oyster Roundtable discussions. He is a hand tonger, fisherman, and hunter. When I called him about going out with him for the day he was quick to offer the invitation, but he wasn't tonging on December 13. "That's the first day of duck season, and I'm going hunting. I keep my priorities straight," he assured me.

I told him that the only other day I could go would be the seventeenth but I couldn't get to the boat until about noon. "That's no problem. Maybe we'll go for some catfish. You can write about that, too," he said sounding like my editor.

I was able to rearrange my schedule for Monday and spoke again with Bob on Sunday evening, December 16. He invited me to go tonging with him. "Meet me at my house about six thirty tomorrow morning. We're going hand tonging."

One of the first things I learned about spending time with Bob Evans is that he enjoys talking. He has worked the water all his life, or at least most of his forty-six years, and he has an opinion based on strong conviction on just about every subject dealing with the Bay, its tributaries, and the life of a working waterman. I would later learn that his opinions are very sound.

At 6:35 A.M. I knocked on the back door of his house. The porch light was on and so were the Christmas lights running along the gutters. Bob yelled, "Come on in," and I entered and was greeted by his two-year-old lab that "has papers a mile long and is a hell of a bird dog." James "Jimbo" Emmert was seated with Bob at the kitchen table; they were having coffee and smoking cigarettes. Jimbo's grandfather held the first soft clam fishing license in Anne Arundel County and, according to Bob, Jimbo "can find some clams—probably the best clammer I know." Jimbo would be working hand tongs with Bob since they were using a boat belonging to Jimbo, who is getting back on the water after leaving for three years to be a private investigator.

Bob told me to take a seat at the table and we talked at length about hunting and dogs. I felt many times that Bob would rather be hunting than working and knew that he didn't oyster before Christmas because duck season was on. I don't think there was really any question where he would be, since he'd rather hunt than do anything else. He and Jimbo talked in detail about dogs' retrieving, hunting techniques, training techniques, and more. As they talked, Bob put on several layers of clothing, and forty-five minutes later, after he threw some sandwiches together (country ham and cheese and crabmeat, no mayonnaise or mustard), we walked out the door.

A handsome, young black man was getting out of Bob's diesel pickup and I was introduced to Amos Jones, nineteen, who has been Bob's crew for five years—crew, that is, for oystering, fishing, hunting, working around the yard, whatever. "He's a nice young man. Hard-working and smart. I don't know what I'd do without him," Bob would say several times during the next two days.

Amos and Jimbo get into Bob's gas truck, and Bob tells me proudly that we're going in the diesel "with over 340,000 miles on her." We're going to Glebe Creek, off the South River in Mayo, just south of Annapolis. As we begin the thirty-minute drive I ask Bob about how it is working the water these days.

"Well, I'm very active on the political side of things, you know. I stay involved in the Maryland Watermen's Association and the Anne Arundel association. You've got to be because there's too much politics in fisheries these days. The CCA [Coastal Conservation Association] hammers us on every issue. They don't want us to catch a thing because they want it for their recreation fishing members. It kills me because now they have coined a new term, 'overexploiting' the resource instead of overharvesting. It's bull, but it's what we have to live with.

"I'll tell you the Bay is in good shape. The fish industry is back, there's some oysters, and while we didn't have many crabs this past summer, we had more than we knew what to do with in the fall. I disagree with most of the fisheries managers and people from groups like the CCA and [Chesapeake] Bay Foundation who are always crying wolf or acting like Chicken Little with the sky falling all around 'em. I don't know why they don't listen to us," he asserts.

"Watermen can't afford to live on the water anymore, and now the new people moving in don't even want us near them because of the noise and our dirty boats. You think I don't know my damn place doesn't look like a junkyard to those people? Well, it may, but I've got to work five or six different fisheries during the year and I use every bit of that stuff to make my livin'.

"Before I bought my place in Churchton, I was making a house payment, a payment for my retail seafood business, and slip fees at Parrish Creek. The storefront rent went up and so did the marina so I was working to pay rich people to get richer. It was bull, so I consolidated and moved everything to the commercial site in Churchton, lock, stock, and barrel."

I asked him about the problem of finding slip space and if it was as significant as others say and as I presumed.

"We're in a hell of a spot, you know. The marinas don't want us, and even if they did we couldn't afford the slip rents they charge in the county. You tried to help us over ten years ago with finding property, remember that, Mick?"

"I do. It was when John Orme was president of the Anne Arundel County Watermen's Association and we looked at the Johns Hopkins University site. But nothing ever happened, did it?" I asked.

"Well, we're still workin' on it. It did sell and now we're trying to see what might happen. This county doesn't care about the watermen—never has. I saw where you wrote that letter to the Maryland Marine Trades Association and the Marina Operators Association [of America] recommending that they work with us. We appreciate that. I don't know what we'll do if something doesn't get worked out, you know," he says.

As we approach the Mayo peninsula, I tell him that I was surprised that there were oysters to hand tong on the South River.

"Well, like a lot of places on the Bay, we hadn't had neither an oyster for fifteen years. Then when those people at Holiday Point Marina wanted to expand into what we said were wetlands near an oyster bar, all hell broke loose."

"I remember that," I acknowledged. "I looked at that property when I was with the MTAM, and while they wanted our help, I told them we couldn't support the expansion if it meant intrusion on wetlands and an oyster bed. They weren't happy and I think they quit the association."

"Well, there were four hearings on that expansion and I testified with the Anne Arundel County Watermen's Association," Bob explained. "When it came time for the final hearing I got a call from some-one at DNR [Maryland Department of Natural Resources], and I won't name names, who asked what it would take to make a deal. I knew the politics and money involved in this marina project's efforts and that we'd probably lose. I think ol' Louis Goldstein [the late comptroller of Mary-land and historically one of the most powerful people in the state—if not the most powerful and influential] was involved, so what chance did we have to fight it?"

"What'd you do?" was my big question.

"I told the state guy we'd back off opposition if they gave us ten thousand seed oysters for ten years. I shocked the crap out of him. He said he'd have to make a few calls and get back to me. I told him to make it before the hearing or it would be too late. Well, later that night, he calls me back and asks if we'd settle for five thousand seed oysters for five years. The deal was done as far as I was concerned and now we have oysters on the river, but I paid a hell of a price for cutting that deal," he said.

"What happened?" I asked wanting more.

"Well, Johnny Orme [then-president of the Anne Arundel County Watermen's Association] was opposed and so were a bunch of the other guys, you know. I got anonymous phone calls at night threatening to burn my house down for not testifying against the marine expansion at the last hearing. God's truth. Problem was some of these guys let the thing get personal. If I've learned one thing from Larry [Simns] over the years it's don't let it get personal. Keep it business and don't show a lot of emotions about it. Cut the right deal for the industry in the long run. I think I did. Listen, when I was a boy I could catch fifteen muskrat a day where the damn marina parking lot is built on wetlands so nobody can tell me I didn't think long and hard about it. Now the guys who did the most complaining are the ones out there tonging on the river today. Ain't that a hell of a note," he concluded.

Jimbo Emmert would later tell me that as far as he and other water-men are concerned all the oysters on the South River should be called "Bob Evans oysters" for what he did that night.

I think back when Larry Simns had to live with a rockfish moratorium established in 1985 to revitalize this famous commodity, and it
lasted five years. Simns and the watermen had fought hard against the
moratorium, but someone considered Simns "guilty of failure" and his
boats were vandalized during the night after the DNR's position was
announced. Perhaps there are things we will never understand about
these men. Clearly, the members of the MWA, like it or not, have had
politically sensitive thinkers who led with an aptitude for looking to the
future. Without them, as Bob Evans speculates, "There might not even
be an association or an industry. And these guys, me included, have paid
many a pretty personal price for some of the decisions we've had to
make. Larry Simns is a true statesman. If I have a question I call him,
and in all my questions over a lot of years his answers have been 100 percent right."

"I guess what worries me," I offer, "is what will happen in the next
few years when Larry retires as president of the MWA. And Betty retires too for that matter."

I can tell by the frown on his face that Bob does not like to think
about these problems.

"I'll tell you. We've thought and talked about this and we have some
guys in mind. Nowadays it's gonna have to be the right man because
we're in a different world now. We can't have someone who's gonna go
in there like a bull in a damn china shop. He's going to have to be
right—young, charismatic, political, you know. It will be a hard transition and even harder thinkin' of what we do with the association stuff
that Betty does: the fundraising, membership, meetings, insurance programs, newspaper, and all that other stuff. We've got to think hard on it,
you know. But its time is comin'," he concludes as we pull into a makeshift parking space at the dock where Jimbo's boat is tied up. Jimbo and
Amos have started the engine on *Lori P,* lit the kerosene heater in the
cabin, stacked up bushel baskets, and made ready to cast off when Bob
and I arrive.

Bob Evans first learned about hand tonging when he was thirteen
years old from his cousin John Sheckells, now seventy-two, who occa-

Bob Evans, Amos Jones, and Jimbo Emmert, *left to right,* are optimistic judging from the number of bushel baskets they are putting aboard.

sionally will go out on *Lori P* with the younger men. "I learned everything Sheckells knows, then added a whole lot to it," Evans says with a smile.

As we move out of Glebe Creek onto the South River and down opposite Harness Creek where several boats should already be working a good bed of oysters, Bob explains that there are a number of things he worries about.

"Pfiesteria, and what that has done to the public relations of selling fish, has hurt. As chairman of the Maryland Seafood Marketing Advisory Commission at the Department of Agriculture, I know it's a big effort to get people to understand about pfiesteria. It's an algae, you know, that's always in the mud, but apparently when it gets stirred up it has twenty-seven stages of life it goes through to turn into this pain-in-the-ass which kills fish or causes lesions. I personally do think it comes from chicken manure running off into the creeks and rivers, and the state and everyone is addressing those issues," he says.

"Another thing is the environmental groups like the Chesapeake Bay Foundation calling for things like an oyster moratorium a few years back or a crab moratorium this past summer. They don't know what they're talking about a lot of the time, but they have such effective, far-reaching PR the public believes that what they say is gospel. We don't have the PR organization, numbers, or money to fight them on this stuff and most of it's to raise membership," he resolves.

Bob and Jimbo talk about identifying someone who can fix them up with an electric power winder to attach to their hand tongs. "The winder would allow us to save our strength and pull up a lot more licks. There's a boy over on the Nanticoke River who has one rigged on his shafts and he can get three licks to our one," Jim explains. "You raise tongs all day and everything burns in your muscles—back, shoulders, legs, and arms."

The shafts on the tongs vary depending on the depth of the water the men are working or their own preference. They have shafts ranging from fourteen to over twenty feet in length. "For me it's better working the long shaft tongs," Bob explains, "because the weight from the wood can balance out with the weight of the oysters raised in the tongs and I can move it okay to the cullin' board. Shorter shaft tongs you've got to bull 'em up to the board."

There are three other boats working the oyster bar when we arrive. Two are open twenty-three-foot SeaHawks; the other is a forty-two-footer like Jimbo's. We're in twelve feet of water and the men are hoping to catch the fifteen-bushel limit well before the three o'clock deadline. They can actually catch thirty bushels because both Bob and Jimbo have licenses, but they will be happy with fifteen. "It's cold and the breeze is comin' strong from the west-northwest so it will be a rough day," Bob says.

Bob and Jimbo each take positions on *Lori P*'s washboards with pairs of fourteen-foot shafts on their hand tongs. The longer shaft tongs are placed on the roof of the cabin so that they are out of the way but easily reached if one of them wants to switch to a different length. The boat has two wooden culling boards, one on each side of the boat just behind the cabin. Amos positions himself on the starboard side behind the cull-

Jimbo Emmert with his hand tongs ready to make another lick. He waits for the wind to turn the boat into a favorable position over an oyster bar.

ing board, his back to Jimbo, and places two bushel baskets at his feet. One basket is for select oysters—large, fat, and juicy—that will be shucked at Bob's house and put in pint and quart jars for retail sales. The other is for the rest of the legal oysters which will be sold at today's market price of $20 wholesale or $39 retail for people who want to shuck their own oysters as I do.

Bob and Jimbo pace back and forth along the washboards like lions in a cage at the zoo. They hold the wooden shafts together with both hands and gently tap the tongs against the bottom. "We're feelin' our way now," Jimbo says. "When we come to a hard rise, that's oysters."

Jimbo is the first to stop, wiggle the tongs against the bottom, and pull them up, hand over fist, until he's about five feet from the tongs. He then mechanically turns toward Amos and the culling board, coming up just to the outside of Amos's right shoulder and drops his catch on the culling board.

Amos, like Kenny Keen, begins to go through the small pile, quickly grabbing oysters and dropping them, without looking, into one of the two orange plastic bushel baskets—one for select, the other for regular oysters.

Bob is still working the bottom as Jimbo brings up his catch again and drops it before Amos. As Bob gets into his rhythm, he begins to drop oysters on the other culling board. Amos moves over and starts the same process of culling Bob's catch and disposing of oysters as he did Jimbo's—select and regular.

"Hand tonging for oysters ain't easy work," Bob yells to me over the drone of the engine. "None of this is, but it's not a job. It's a way of life and there's a big difference." That's all he has to say about that, as he turns to pace toward the aft section of the boat.

In twenty-degree temperature, with a windchill well below that, my hands are starting to freeze. I go into the cabin and grab a pair of cotton gloves to borrow for the remainder of the day.

Whether using patent tong rigs or dredging from a skipjack, oystermen randomly pull up whatever is on the bottom within the bounds of the oyster bed. With hand tongs, oystermen feel the bottom with their tongs until they hit a lick. For divers, success comes from physically

Amos Jones culls oysters as Jimbo Emmert empties his hand tongs onto the culling board.

touching the bed with their hands. In feeling the bottom Bob comments that "it's weird because it's hard as a rock like a pile of oysters but there aren't any oysters on it. With this boat swinging back and forth in this wind I can't get a good lick."

Unlike working the patent tong rigs in the open Bay, the hand tongers work on smaller bars and talk back and forth between the boats while they are working. Bob's radio isn't turned on. He yells to some boys from St. Mary's County, "I sure would like bein' home with my feet up eatin' one of those St. Mary's County country ham sandwiches instead of tongin' these oysters."

"Ain't that the truth," was the response carried across the river.

With the tide moving quickly, a man on one of the SeaHawks puts his boot against the shaft as he dips it in the water to make sure it goes straight down. Bob stated that a "heavy tide in a stiff breeze like this makes it hard to walk the boards, work the bottom, and lift the fifty-to-sixty-pound tongs with each lick when she's shearing off in one direction or another. The anchor doesn't keep her straight when it's this windy. I hate tonging like this." He steps down and shifts the anchor from port to starboard. "Just when we think we're on a set of oysters the boat moves. Let's see if this helps." As he's moving the anchor line he notices a boy on the SeaHawk down on his knees on the washboard. "First time I've ever seen anyone tongin' on their knees," he yells to the boat. "He looks like he oysterin' for Jesus today." The men laugh and carry on.

There are six boats working this one bar now. We see a small jonboat coming from across the South River. When we can clearly see the dogs pacing in the bow, we realize the men aboard were duck hunting. The boat comes up and stops to talk with John Orme on *Traveler* who is working approximately fifty yards from *Lori P.* After spending a few minutes with John, the boys head toward us and I recognize John Flood, a local environmental activist and Rotary member. He holds up a couple of ducks as I take their picture and he says, "We're authentic Chesapeake Bay duck hunters, Mick." His companion this day, Bob explains, is "a boy named Drew who works for DNR inspecting pound nets."

In 1988 over 90 percent of the Bay's oyster population was killed by two parasites, MSX and Dermo, aided previously by the devastating hurricane Agnes that blew up the Bay and left destruction of oyster and clam beds in its wake. During the past decade, DNR officials and watermen have been working in partnership to plant seed oysters in areas they think will produce good oysters while avoiding an invasion of MSX or Dermo. "We don't always agree with the state on where to plant the seed," says Bob, "but they call the shots. Now you take this river. We didn't have neither an oyster fifteen years ago. Then me and Kenny Keen planted seed for the state and we got good oysters growing here now. We worked a bar on the other side last year and this summer they had a good growing spurt to three-to-four-inch oysters, then Dermo killed over 60 percent of them. These on this side are all right. Dermo doesn't kill an oyster until it's matured to three or four inches so it's a bitch 'cause that would have been another good year on that bar—maybe a bushel an hour—but it's gone now."

The wind has picked up to twenty to twenty-five miles per hour, making for whitecaps and strong currents pushing water up and over the transom. Adding to the conditions is the temperature, which now stands at about twenty degrees without factoring in the windchill.

A stack of four empty wooden bushel baskets is blown overboard from one of the SeaHawks. The crew tries to retrieve them with boat hooks, but they are soon lost to the South River. Amos had been tonging while Jimbo and Bob culled but was down off the washboard within ten minutes. "I ain't goin' overboard for a handful of oysters. That wind grabs the shafts and damn near throws you over." Bob tells Amos and Jimbo to lay up the tongs as he wants to move to a bar down river, off Hillsmere Shores and Duvall Creek.

"We'll move down there where we can use that point and work to lee of the wind. We'll catch more oysters in calmer water if we can find them," he says.

Bob can't quite remember exactly where he planted the seed several years ago, so in order to find the bed, he takes a three-foot section of heavy chain tied to a thirty-foot line and drops it overboard in eight to ten feet of water. As Jimbo drives the boat, Bob feels the chain moving across the bottom and can locate the bed, an area where the chain will move in bumps and rises once we're over it. When he says he hit oysters and where the chain drops off the bar, the anchor goes overboard to steady the boat. Amos throws overboard an empty plastic Texaco oil bottle tied to a short line, to use as a buoy marker.

Jimbo climbs up on the washboard; he says he wants to use the fourteen-foot shaft tongs and Amos agrees, putting his eighteen-footer up on top of the cabin and pulling down a shorter pair.

Bob culls the few oysters Jimbo and Amos drop on the two culling boards and says, "Everything is political to me whether it's oyster recovery, legislation, regulation, or keeping our piece of the pie. Watermen don't want everything out here but we need to make sure we keep our fair share, and it gets harder every year."

In the unusually quiet and tranquil area off Hillsmere Shores I watch Amos and think about Vince Leggett and his project to write a book and produce other information about the history of black watermen on the Bay. Amos says he loves working the water and if Bob's character analysis is correct, there will be another good waterman working the water in a few years. (In fact, Amos went out on his own nine months after this trip.)

After thirty minutes and half a bushel of oysters ($10 worth) Bob is convinced he missed the bed and starts dragging the chain again. Two

Hand tongs with shafts of various lengths rest on the culling board. The depth of the water will determine which set to use.

bufflehead ducks land by us, and there are over fifty swans seventy-five yards away on the beach at their Hillsmere Shores winter grounds watching us out of boredom. Bob pulls up his chain and tells us that we're done for the day. It's blowing too hard and they can't "hit on anything. Besides, it's too damn cold to stand out here and hunt for oysters," he says.

Before heading in Bob wants to look for the seed bags he and Kenny Keen planted three years earlier. He tells Jimbo to drive down to beacon 6 toward the mouth of the river, turn around and bear on the next upriver beacon, 8. The seed bags, he claims, should be just outside the clam bottom and near a dive bottom. They were planted in seven feet of water. The depth finder shows thirteen feet as Jimbo heads for beacon 8. "If those oysters were planted here, somebody moved the bottom," Jimbo shouts at Bob who ignores the jab. "You got your beds mixed up. I bet they were on the other side of that beacon 6," he continues. Not hitting a bed in this search, Bob pulls in the chain and says for Jimbo to take us home. It's only one thirty and we have four bushels of regular oysters, plus almost a bushel of selects, which will be shucked and put in pint jars.

"I generally get about twenty-five oysters to the pint. You buy 'em in the store and they'll put thirty in a quart. The rest is water," Bob says parting with more information. Looking at our catch for the several hours we were out, I remember that in 1998, 71 percent of oysters were harvested by hand tongers, 14 percent by divers, 10 percent by patent tongers, and only 5 percent by skipjacks. Tomorrow will be a better day.

As we move into Glebe Creek, Bob hands me a sandwich, crabmeat between two slices of bread. Amos and Jimbo eat ham and cheese. Bob has one of each. After eating, Amos goes outside the cabin to wash the boat down and organize the bushel baskets and tongs. He's making ready to unload the boat so that we don't have to spend much time at the dock. Pulling up to the dock at his private pier in Mayo, Bob shouts, "I'll be damned, a crow with a white wing."

Amos bends over laughing as he states, "A seagull shit on it." We end a rough day on the water with a good laugh.

Back at Bob's house, the oysters are unloaded and we're standing by his truck when waterman Charlie Quade pulls up. He has been laid up with a serious back injury, courtesy of overenthusiastic hand tonging two weeks ago. Exchanging formalities with everyone, he asks Bob if we were oystering or catfishing today. Bob tells him we were tonging on the South River but it was a lousy day with too much wind. Charlie says that the last month "was so easy tonging on that river it was like takin' candy from a baby." Bob chooses to ignore the analogy.

Talk moves like machine-gun fire covering topics ranging from the electric winder to its not being cold enough for Charlie to go perch fishing. Hunting is always a favorite topic for Bob and Jimbo so Charlie proceeds to tell us about his first deer hunting trip to Virginia where he "used dogs and shot buckshot." He recounts that at a dinner the night before the hunt he listened to about fifty or sixty hunters talk about their

As the season gets underway, thoughts of oysters and daily quotas are floating subconsciously through every waterman's mind.

experiences with buckshot. They left an indelible mark on his mind that night that a buck might get up several times after being shot with buckshot. "When I saw this buck I shot him, and then when he moved I shot him again. His antlers looked like Swiss cheese." We were all laughing as Charlie told his stories.

Bob asked Charlie when his boat would be ready because the same man was working on Bob's boat. "Should be ready next week, by the time my back is all right. But he's been down and not able to work lately either," Charlie says.

"I'll tell you what. That man has a case of 'lightitis.' He's been down with Bud Light! He had this much to go [he holds up his right hand indicating a space about an inch wide between his thumb and forefinger] on my boat before she was finished but I don't think he's taken a sober breath since," he says as laughter spreads all around.

I head for home at about three o'clock as the men go inside for a cup of hot coffee.

SKIPJACKS AND OYSTERS

Historically speaking, the uniqueness of the Chesapeake Bay brings to mind the image of the sailing skipjacks, those old-timers of the working fleet, dredging oysters under sail, loaded down to the gunnels with this treasure of the Bay.

Sadly, with the demise of the oyster fishery there are fewer and fewer of the skipjacks working the Bay. "Some of them aren't seaworthy enough to get out in the open water," according to Pete Jensen of Maryland's Department of Natural Resources. Nevertheless, ten of these magnificent vessels are on the water, many running charters for educational purposes through the Chesapeake Bay Foundation, Echo Hill Outdoor School, or on their own.

Captain Jackie Russell of St. George Island in southern Maryland talked to me in September about oystering, "Makes no sense working her *[Dee of St. Mary's]* because you can't make near a cent off oysterin' down here in the southern part. I'll go out with my education program in March but she's not goin' anywhere this winter."

When I asked him about clams he said, "Neither a soul workin' them down here either. You can't find a dozen with a magnifying glass, you know."

With Jackie off the water there were only the few Deal Island boys and probably Russell Dize and Wade Murphy from Tilghman dredging. I told him I'd call Russell because Murphy's *Rebecca T. Ruark,* the oldest skipjack in the fleet at 113 years, may still be laid up.

Murphy and his crew had had a fair day oystering at the start of the season, dredging about seventy bushels, when at three o'clock the seas were churning and there were winds of sixty miles per hour. *Rebecca T. Ruark* foundered in twenty feet of water, spilling both crew and oysters into the fifty-degree water of the Bay's Choptank River.

Speaking to a packed house at the Salem Avery Watermen's Museum in Shady Side, Maryland, Murphy said, "We finished working in the protected waters of the Choptank, pulled the yawl boat [a small motorized push boat allowed by Maryland law to push the skipjacks to their destination where they dredge under sail], and started sailing for Tilghman. I've done it hundreds of times over the past forty years.

"We were doin' okay, even in sixty-mile-per-hour winds, until it started blowing the sails away. It blowed the sails away, and it was the biggest sea I've ever seen in my life," he continued. "I dropped the anchor; the anchor held her to the wind. Then it started breezin' up more. She started diving her bow under. Sometimes there'd be water up on my forward deck up to my waist. My crew was scared to death."

Two of his three crewmembers were brothers who had been with Murphy for three years. "Now both those guys say they'll never spend another day on a skipjack," he said.

Murphy called his wife on a cellular phone and told her where he was and that someone had to meet him and tow him into Tilghman. Wade Murphy has been a waterman for forty-three years. He is known and respected Baywide for his tenacity, but this day tested his grit to the core.

"I told the boys to keep the water out of the boat by keeping the two thirty-five-gallon pumps running. Problem was I didn't know they weren't checking the pumps because they were so scared. They said the water was gaining over the pumps and I told 'em they better start bailing 'cause if they didn't she was going to sink.

"As the watermen arrived and began towing this good boat to shore, she took so much water over the bow, she went down," Murphy said. "I couldn't believe she was goin' down. She started down and I thought she was going to come back up. She never did and I got off her. I made it to this life ring, and about that time, one of my crew, a big fat boy, popped up alongside of me and grabbed the ring, and I said, 'Shoot,' " he explains, emphasizing his new predicament as laughter flows through the audience.

The rescue boats were trying to get clear of the sinking skipjack. "The boy who couldn't swim was so scared he was holding onto the last inch of boat. They threw him a life ring and he wouldn't take it. The third time, they were screaming and finally they pulled him on the boat [just when] I thought for sure he was going to get drowned."

With young Jason Wilson on *The Island Girl* the watermen saved their colleagues from certain disaster in waves remaining at ten to twelve feet. Safe on board, Murphy began to think about raising his hundred-year-old vessel. While he recognized that dredging oysters is not the only way to harvest them, it is the hardest, and Murphy was a committed loyalist to this traditional method.

After calling two commercial companies to raise his boat he found that the cost was going to exceed $30,000. He had spent $80,000 in 1994 to get her into shape but now, five years later, he was thinking, "It was all up."

But in this close-knit society, competing for limited fisheries from sunup to sundown, often in bad weather, there is also loyalty to each other when times are hard. Wade Murphy was not without friends.

"They had a meeting that night on the island and a lot of people were there. I have to give Levin [Captain Buddy] Harrison of Harrison's

Chesapeake House credit. He told the people that this boat was sunk, and if she wasn't gotten up in a day or so, she was going to wash up on the beach," he said. "This was the oldest boat in the Chesapeake and they ought to try and save it, he said. They told me that within two hours, the governor okayed a crane to come from Baltimore."

Harrison's message leapfrogged from the Maryland Department of Business and Economic Development to the governor's Office of Business Advocacy to the Maryland Port Administration. Three days later the state committed $10–12,000 to Baltimore's Martin Imbach Company to raise her. The job was done with great fanfare and concern; major articles appeared in the *Washington Post* and *Baltimore Sun.* The eventual cost to fix her will be about $80,000, and fund-raising efforts are underway. This includes a new Save our Skipjacks Task Force created by Governor Glendening. This task force made quick recommendations for a Skipjack Insurance Trust Fund, reserved oyster beds for skipjack dredging, created state work for skipjacks reseeding oysters, and more.

There was a time when skipjacks dominated the oyster fleet working the Chesapeake Bay. In the mid-to-late-nineteenth century, hundreds of boats were dredging under sail, as a result of the 1865 law passed in Maryland restricting the harvesting of oysters to sail power. A record harvest of fifteen million oysters in 1884 saw nearly a thousand oyster licenses issued that year, for harvesting from various types of sailing vessels. The skipjack was born around 1891, and because it had a flat bottom and straight sides, any reasonably skilled backyard boatbuilder could build one. Soon, there were over six hundred of these new boats working the Bay. However, following the decline of the oyster harvest after its peak in 1884, only ten working skipjacks remain at the beginning of the twenty-first century. That small group is America's last fleet at work under sail.

It will take at least the proverbial village to restore the Bay's oyster fishery and the ecological fabric of the watershed. While most people know oysters for their culinary attributes, perhaps their true value, from a Bay perspective, rests in ecology. Oysters, as bivalves, filter water for their food. Dirt, nutrients, and algae are either eaten or shaped into small packets and deposited on the Bay floor where they are not harmful. Consider that oysters in the Bay could once filter the entire body of water in three to six days. The same job would take the Bay's remaining oyster population almost a year. In addition, the hundreds of animals who use oyster beds for habitat, from grass shrimp to mud crabs, provide a food source for striped bass, weakfish, black drum, croakers, and the popular blue crab. In the nineteenth century, oyster reefs—and they were reefs—were so large that they were considered navigational hazards. After 120 years of intensive harvest, very few reefs remain. People call them beds because they are flat rather than mountainous.

During the latter part of January and the first two weeks of February, ice covered most of the Bay and particularly her tributaries. Boats were frozen in, which meant that few men could get out on the water. By mid-February the Maryland Tidal Fish Advisory Committee and Department of Natural Resources Secretary Sarah Taylor-Rogers announced an extension of the oyster season. The committee had voted unanimously to extend the season from March 31 to April 14. Taylor-Rogers pointed out that the men had not been able to dredge for at least two weeks, and with a strong oyster market and the good health of the oyster population the state wanted to capitalize on the opportunity.

While boats were trapped up the tributaries in creeks of ice and fishing remained quiet from the upper Bay to southern Maryland, there were still a few hardy souls, like Rock Hall gillnetters Danny Elburn and Chris Lingerman, who could get out of the harbor. A waterman at one of the quiet Chesapeake Bay gillnet striped bass check-in stations knew of the one harbor that was not frozen and commented on the work ethic of most watermen: "Some of those Rock Hall boys must be fishin'," and so they were.

"I've been going out pretty near every day," Lingerman told the *Waterman's Gazette*. "When all is said and done today, we've probably got a thousand fish." And, according to Lingerman, who started working at two thirty in the morning, another good snow would enable him to catch even more fish before other men could get out on the water. "Snow pushes fish to the bottom where the nets are."

While snow and bitter cold make for better fishing for a few men fortunate enough to have access to open water, others are trapped in ice and operations are stalled for weeks on end. When the weather warms and the first boat can break through the remaining crust of thin, gray ice, watermen will be leaving harbors and marinas like wagon trains going west. I would make plans to be on several of the boats.

CHAPTER 3

Fisheries and the Future

Between Christmas and New Year's Day, Bob Evans calls to tell me he's going catfishing and asks if I'd like to tag along. "Definitely," I respond enthusiastically and agree to meet at his house at nine o'clock the following morning.

Catfishing is already pleasing to me because we don't have to go out on the water in the frigid early morning darkness. While it is cold and overcast today, it is still nine o'clock and I'm looking at the sun making a weak attempt to push through a layer of winter cloud cover.

When I arrive, the activity at Bob's house is quietly buzzing like an office in the early morning. He is sitting at the kitchen table smoking a cigarette and talking on the telephone or with Jimbo and/or Amos. You get your own coffee and are welcome to it. His daughter and the dogs are drifting through the morning ignoring Bob's conversations. When he is ready, Bob gets up from his chair, moves to the counter, and literally slaps together sandwiches of country ham and cheese or crab and codfish. He fills thermoses with coffee and asks Amos if he has sodas (Amos doesn't drink coffee). After tossing the food into a small Coleman cooler, he moves to the closet to begin putting on several layers of clothing, two pairs of socks, and his white rubber boots, a waterman's trademark. As he's dressing he is giving

Amos instructions for the day regarding the truck, the bait, the gear, clothing, or other items from a laundry list in his head. Jimbo moves from the table, putting out his cigarette. I sense Bob is nearly ready and the body language says we better all get a move on. It seems that as long as he is sitting at the table and keeping the conversation running like a fast-moving stream, everyone is relaxed and in the flow of the discussions. However, once the first phrase specifically directed at the day's harvest activity is uttered, or he moves from the chair to the closet, the informality is over and things get serious. It's time to go to work!

As we walk to the trucks, we see that Amos has already loaded three hoop nets in the truck he will drive with Jimbo. Hoop nets are actually like traps, but the men still call them pots. They consist of a set of fiberglass hoops or rings approximately 3½ feet in diameter by 8 feet long with two interior chambers. The mesh size is 4 inches in the front of the net and 3 inches in the tail. PVC pipe rods run the length of the nets and keep the net stretched out straight. At the end is a drawstring that is used to open and close the net. Once the fish are inside they can't get out. The rings are encircled with rope mesh that appears shiny and black. I asked Bob why they look this way.

"We dip the nets in tar before we make 'em up. Watermen all over the Bay have been using tar to preserve their nets for decades. The tar preserves them better than anything else does. They will stay in the water from November to June and that can tear up a rope net, you know. We generally retar them in the spring. Me and Danny Beck [a fisherman from Middle River, Maryland] go in together and buy six hundred gallons at a wholesale price so it ain't too bad."

As Bob and I get into the diesel truck, Jimbo comes over to ask Bob if he has talked to anyone about the electric winder rig for the hand tongs. Bob says he talked to a couple of men last night but no one was making any up. "We'll make our own by jury-riggin' a hydraulic setup for the hand tongs," Bob says as he turns over the diesel. He's tired of searching and asking so he tells me they can "put some stuff together and make it work somehow." I believe they will.

Stopping to get fuel and cigarettes at Park's Liquor in Deale, Bob says, "These fuel prices are sure eatin' into my pocket." I ask him what fuel is running him, and he says he has already spent $90 this week and it's only Tuesday.

As we drive toward the upper reaches of the Patuxent River where freshwater and catfish are found, I listen as Bob goes into a dialogue on bass fishermen. "Bass fishermen are fanatics. Don't want neither a soul fishin' the river. Why a few years ago a boy came down from upper Bay and, you see, you always lose a few cats in the cribs [live boxes]. They die from changes in temperature or whatever. Anyway this boy releases about two hundred pounds in the river and they wash up on shore. To the rich people who own the waterfront that looks like a damn fish kill and before you know it we're starin' at possible legislation to ban fishing on the Patuxent. I had to go down and get that straightened out. See that boy was used to workin' up the Bay where there's a lot of waterfront still owned by watermen or people who understand our needs, you know. But not down here on the Patuxent any more.

"Anyway, it's the same damn thing with the bass fishermen. They don't want us on the river, and they have sabotaged my gear and nets. Every other year it seems they hit me. They even go to seminars on how to destroy commercial fishing gear. Now that may sound like bull, but I know it's true. They turn fish loose, cut off pound net poles, cut our nets, and steal them. Hell, they even stole nets that belonged to the DNR perch survey! Truth is, we don't have problems until we get up in the freshwater for catfish, which is also bass territory."

Although he concentrates on hoop net fishing, he can also use the fyke net, which is cylindrical, similar to a hoop net or eel pot, only larger and with a hard frame so that it can be anchored in position or secured to stakes. "The bass fishermen don't like the fykes because we use them for white and yellow perch, too, and we might catch up to twenty different varieties in a day, including bass. I release what I'm not harvesting but they don't care. They don't want me fishing, period. And one more thing. I'll tell you that I don't mess with any recreational fishermen when they're fishing, and I start catfishing later in the day because I don't want to upset the duck hunters. They get pissed if we start workin' nets and they're workin' their decoys and dogs from a blind or the marsh. I leave 'em be and wait until they should be off the river. Point is, in reality, commercial fishermen don't get treated with the same courtesy as we treat other users—and that's a fact!"

Thinking about catfish I ask Bob about eel fishing. "I used to fish for eel in the spring when they started to move from the mud. They eat the catfish bait. I'll tell you what, in the 1970s there were so many eels it would mesmerize you. I'd make $7–8,000 a week in my heyday. Trouble was the mesh sizes weren't regulated so men caught up all the little eels. Hell, I gave it up ten years ago because the fuel was costing me more than I made. There's boys eeling because of the overseas market and all, but I don't fool with it anymore," he says.

Even though we're going catfishing, Bob is on a roll now. "I've talked a little about bass fishermen, eels, and now let me tell you somethin' about the crab industry," he states emphatically.

"Go ahead," I say with mild enthusiasm because I know I can't stop him and I'm interested in his perspective. I've found that most people who get caught up on a subject near and dear to their livelihood make statements that are mixed with emotion, but what they say is about as close to the truth as you can get, despite the fact that it is somewhat tainted with exaggeration.

"You know Phillips Seafood, the restaurants and all?" he asks and I assure him that I do. "Well, Phillips is about to monopolize the crab-meat industry in the country. This fall they cost me five to eight dollars a bushel because the East Coast packers cut their prices. My females go to New York for the oriental market, as well as to California and Canada. We have a nice, fat female, you see, and it can bring good money but they only brought twelve dollars a bushel. It's pathetic. And number ones [large male hard crabs] were only sixteen dollars a basket. You see, the packinghouses couldn't keep up with the crabs. There were so many, so the prices went down.

"If we get twelve dollars a bushel and we have a thirty-bushel limit on the Potomac, we can't make it on $360 when it costs us $200 to leave

the damn dock. We didn't have neither crab all summer and then in the
fall they came out of the woodwork. You can't figure out crabs. But I'll
tell you, we've got to create a fall market. I was selling a hundred bushel a
week last fall from my house, and I'm getting $45 a bushel instead of $16
from the packinghouse," he says. I make a note to talk more about this
later in the book.

While I had heard Bay watermen talking for years about labeling
crab meat and concern over imported meat, I didn't give the Phillips
crab issue much thought until Delegate Richard D'Amato, a Democrat
from Annapolis, introduced a bill in the Maryland legislature requiring
the state to design and require uniform and prominent labels to define
the place of origin of crabmeat. Violators, according to the law, would
be charged $5,000 per offense.

"I believe Marylanders would choose Maryland crabmeat because
the taste is sweeter and fresher than Asian imports and out of pure loyalty to our watermen. Consumers should be given a choice," D'Amato
said.

Phillips Food, which owns and operates the popular Phillips Seafood restaurants, including one in Annapolis, says the bill is unfair and
may run afoul of federal law.

D'Amato cited labels such as "Maryland-style crabcakes," "Chesapeake Bay Brand Crabcakes," or "Product of Maryland" as examples of
what he says is false advertising. "They do not mean Maryland crab. It
means at best that a Maryland recipe may have been used. Chances are
you're eating Asian crabmeat caught and picked by cheap labor," he
said.

What is particularly difficult about this issue is the fact that Phillips
packaged over 3.9 million pounds of crabmeat in 1999, not counting
what it sold in its restaurants. That is nearly three times the amount of
crabmeat produced in the entire state in a season. All the watermen in
the state together brought in only about 1.9 million pounds of crabmeat
last year, according to Mark Sneed, president of Phillips Food. "Even if
I bought every crab that was sold in Maryland, I would still have to buy
meat from Asia," he said. To meet this demand from its customers,

Phillips produces roughly 80,000 crabcakes a day and owns crab companies in Thailand, the Philippines, and Indonesia.

Larry Simns, speaking for the Maryland Watermen's Association, agrees with Bob Evans and attributes the decrease in harvest in 1999 not to a depletion in the crab stock but to a lower market share.

"All we want is a fair shake," Simns says. "We can't compete with the price of the imports. But I believe people would pay a higher price for crab if they were guaranteed it was from Maryland."

Bill Woodfield, head of the Chesapeake Bay Seafood Industries Association in Galesville, Maryland, thinks the situation is becoming critical. "The tonnages of imported crabmeat from Asia are still increasing, more domestic markets have been lost, prices of our products have gone down along with the price of crabs to watermen," he said.

A University of Maryland study put a value of $4.45 per pound on imported crabmeat compared to $11.97 on domestic. While D'Amato insists that he doesn't want jobs cut because of his legislation, Phillips Food's Sneed says that he has created over three hundred jobs in Maryland. He feels that the passage of this bill could mean that Phillips does not need to be a Maryland-based company any longer. As the issue of more than watermen's jobs enters the equation, I, like many others, anticipate a very interesting fight in the legislature over this issue.

Bob and I drive toward the river, and I think about his concerns over Asian imports versus Maryland blue crab. I'm reminded of a visit I made years ago to Pittsburgh. My aunt Jane Blackistone Hughes had invited me to visit the unfamiliar upper reaches of society by having lunch at the exclusive club, Ligonier, which was established by the famous Mellon family. As I scanned the menu of pheasant and other delicacies to which I was totally unaccustomed, I spotted soft crabs on the list and immediately felt more comfortable. I began to ask the chef for the soft crabs when good ol' outspoken Aunt Jane, in front of God and country, announced with native Potomac River conviction, "Don't get those crabs. They aren't from Maryland and they don't know how to fix them up here." I had learned long ago that such a statement by this matriarch was not a recommendation or suggestion, but an order of sorts, born out of love for Maryland crab and for me. To her, nothing could compare, so why settle for less. I tried the pheasant—much to my regret.

It's midmorning when we reach the headwaters of the Patuxent River, beyond the strong influx of salt water, where Evans has placed his nets. We drive down a steep hill to a makeshift ramp and ad hoc parking lot; Amos and Jimbo are already there unloading the bait and gear that will be needed. Bob tells me, "We don't come down here too early because of the duck hunters. I wait until they're gone. One less problem for both of us."

I gaze out over the river with mixed feelings of tranquility and subtle enthusiasm. A lone brown eagle rides the breeze right down the middle of the river. Its dark brown feathers reflect off the calm gray surface presenting a pleasant complement to the light brown marsh grass across the water.

The peace is broken by the men unloading sixteen five-gallon buckets containing a bloody mixture of chum and menhaden parts. "The cats like menhaden now, or ol' mud shad, but then all of a sudden when it gets colder they'll want soft belly clams or horseshoe crabs. They get fickle and you just have to figure out what they want to eat and when. I've got to go to Delaware to get the clams and the same for the horseshoe crabs. I used to catch my own horseshoe crabs for fishing but now with new regulations I've got to go all the way to the Delaware Bay for 'em. And on top of that I've got to pay a dollar a piece for 'em because you can't figure out what the bastards are going to want to eat next," Bob states as he moves down to the water's edge to retrieve a beat-up fiberglass skiff.

He unwraps the line from its piling and pulls the boat over to the ramp. Jimbo and Amos each grab a side of the bow and pull it up the ramp as close as they can get to the bait buckets that have to be loaded. As they do this, Bob climbs aboard with a gas tank and does his best to start a cantankerous old Evinrude outboard engine.

With the engine finally started, a spare five-gallon tank is put aboard. Bob tells me that he, Amos, and Jimbo need to run across the river to retrieve a live box, which holds catfish from the past week's catch. I watch as they move across the river; the only thing breaking the silence is the old Evinrude. I'm pleased to sit in the cold and enjoy the solitude until they return. After fifteen minutes I'm feeling the bitter wind blow off the river and turn to stare at forty square plastic tubs that will need to be filled with fifty pounds each of catfish today if Bob is going to meet suppliers' demands for fish. A buyer in New York wants 1,200 pounds and one in Jessup, Maryland, wants 1,000 pounds tomorrow morning. Bob will be paid forty cents a pound.

When they return to shore pulling the large live box, Bob steers the twenty-foot open boat up on the ramp as Amos directs the box so that it shares the space with the boat. Bob then uses a scoop net with a hoop about twenty-four inches in diameter to begin transferring fish from the box to the plastic tubs. As the tubs are filled, Amos and Jimbo transfer them from the ramp to the back of the pickup. Tub after tub is moved to the truck. "Pretty, ain't they?" Bob says to me as he scoops out a couple of large cats each weighing about seven pounds. The fish will be graded by size and weight before being shipped to the buyers.

Catfish pots await transport up the Patuxent River by Bob Evans and his crew.

I ask Bob about the fishing and he explains that he's taken "a quarter million pounds of fish out of this river a year for fifteen years. In May I took out 63,000 pounds. There aren't a lot of boys catfishing in the winter so it's good for me. But, I hold off on catfishin' for a few weeks in late December because if I don't have an order for at least 1,000 pounds it ain't worth my time to fool with it, and who wants to eat catfish over the holidays? The market stays down during the holidays but it's getting back up now that it's approaching the first of the year." I conclude that just about all decisions are based on weather and market.

After the live box is emptied, it needs to be hauled back across the river. The men make ready for the brief trip, but Bob is having a difficult time with the Evinrude. "This motor is being a pain today," he tells all who are listening as they sit silently in the boat.

Leaving the dock to begin pulling, rebaiting, and placing new pots, Bob points to the shore and says the marsh used to come out over a hundred yards. This means a few years ago we would now be sitting in the middle of the marsh instead of moving at six knots in five feet of water. "Nor'westers came and ate the marsh away," he says with some disappointment.

We approach a buoy floating upriver about fifty yards from shore, and Bob slows the boat. Amos reaches for his boat hook, and when we pull alongside the buoy, he reaches below the surface and hooks a line tied to the buoy. The line is perhaps twelve hundred feet long. Bob steers the boat parallel to the line which is moving through the open end of the hook, and a pot surfaces. Catfish pots are attached to the line with hooks about every 120 feet. The line is anchored to the bottom with the help of an old crankshaft. As Amos pulls in each pot he rests it on the side of the boat, unties the open end, and dumps the fish on the bottom of the boat. While Jimbo sorts fish by size, throwing them into the plastic tubs, Amos takes a five-gallon bucket of clam and fish bait, which smells terrible, and dumps about half of it in the pot. He then reties it shut and drops it overboard. Bob explains that these pots draw fish from a long way up the river so he will spread them out during periods of slow fishing like today.

As we move upriver two miles we leave more of his buoys in our wake. We're traveling at about fifteen knots. I'm not sure what the windchill factor would be when you take into account the movement of the boat upriver against the wind and a temperature below freezing, but all four of us feel the sting of cold against the only skin showing—cheeks and foreheads. We slow down, thank God, at an isolated buoy sitting in ten feet of water. It holds one pot on a short tether. Amos pulls it, empties its contents on the deck, rebaits it, and drops it overboard. The process takes about five minutes and as Jimbo sorts fish and tosses old bait (including fish heads and clam shells) as well as undersized or unwanted fish, we're moving upriver again at a decent clip. This same routine will continue for the next hour. I think momentarily about the constant complaint of environmentalists and some recreational boaters about the commercial bycatch issue and realize there is no bycatch in this fishery. "It's clean as a whistle," Jimbo states as he takes a long draw off a Marlboro Light that he managed to light before Bob opened up the engine between pots.

We are the only boat on the river now that the duck hunters have retreated to warm trucks and hot coffee. More men will work the river in the spring and summer. "I like working the river in the winter if we have market. We just have to be more strategic on where we plant the pots. You move 'em ten feet one way or another and you won't catch fish as well," Bob says as we head back toward the ramp with about four hundred pounds of fish in the boat. Amos uses this ten-minute trip back to the boat ramp and trucks as an opportunity to relax out in the open wind as he stretches across the bow. He takes leisurely drags off a Kool cigarette that he keeps cradled in his palm to protect it from the wind. Jimbo and Bob smoke too, and all of us are staring out over the river as we pass brown marsh, covered by a thin shroud of snow, its icy fingers reaching out into the river for us. In a month this river will be frozen solid.

We arrive at the ramp which is wet and covered with the deep green of algae clinging to the wood. Water temperatures are in the low forties, so Bob warns everyone to be careful getting out of the boat. With over three hundred pounds of fish in six tubs to unload he doesn't want anyone to slip and risk the loss of part of the catch. Minutes later while reaching for a tub, Jimbo's feet slide out from under him and he falls to the ramp. As everyone laughs Jimbo is lying on the ramp in water that comes up to his waist. It's a good time to be covered in oilskins. "You all finished laughin' yet?" Jimbo asks in embarrassed disgust.

"Hey, when the chips are down, we'll be here to grind them into you," Bob replies.

When the boat is unloaded and we have transported all the tubs to the pickup, Bob says they need to make another run down the river to check more pots and put several new ones overboard. That will make a total of eighteen hoop net pots that he wants to leave baited in the water.

We head directly across the Patuxent and slip back into a cove at about fifteen knots. The wind cuts through layers of clothing as the temperature begins to drop because of increasing cloud cover. Slowing the

Bob Evans, *left*, uses a dip net to transfer catfish from a live box into a bushel basket. Jimbo Emmert, *right*, will add the basket to the others on a truck bound for market.

engine to pull up alongside a buoy, Bob tells us that the river slopes off to thirty feet deep and that there are some holes as deep as a couple hundred feet. If he lays his pots in thirty feet of water he will get the bigger cats; fifteen feet will harvest smaller ones. After pulling several pots and replacing them fully baited, we have about two hundred pounds of fish to take back to the ramp and the waiting trucks.

After a quick transfer job, Amos and Jimbo put into the boat three five-gallon bait buckets and new hoop net pots, black with new tar that reflects the winter sun. They are going farther downriver to finish emptying the remaining pots. I decide to stay ashore and sit in the truck to try to warm up. I'm a little hesitant to remain behind but it doesn't faze the men at all. "Turn on the diesel and get warm. We'll be back in half an hour or so," Bob says, encouraging me to follow my instincts.

When I hear the boat approaching I leave the warm truck reluctantly and meet the cold and windblown men at the ramp to help them unload the tubs they've just filled. After we've finished Bob says he wants to pull an eel live box onto the shore to let it dry out—it has been sitting in about ten inches of water. He wants all of us to lift the 6-foot by 8-foot box out of the water and prop it against the bulkhead. With low tide we can maneuver around the box, square ourselves, and lift it upright. As we do so, with two men on each side, at least two dozen live eels squirm out of the box and wriggle in the sand. "Tough, ain't they?" Bob says. "They have been in this box since last spring livin' off algae and whatever else they can get in there." He asks Amos if he wants to take a few to his grandfather but the offer is politely and quickly declined, so we release the eels into the water.

After putting the eel box on dry land for later use, everyone lights up a cigarette and we begin walking along the beach that is about two feet wide from tide to marsh. Amos points out a muskrat run that Bob decides to investigate. He trips in the rat hole, and we all watch helplessly as his two-hundred-plus-pound frame in orange oilskins falls forward into the marsh. When we are sure he is unhurt, laughter echoes across the river. Amos is doubled over and clutching his stomach, crying tears of laughter. I'm so busy watching Bob that I trip over a dead branch pro-

A live box rests on the beach during the winter. It will be used for crabs come summer. At ramps or in yards around the Bay, gear used for harvests will be left to decorate the landscape until it can be used again the next season.

truding three feet from the marsh and fall sideways into six inches of water. The laughter erupts again as Bob joins in now that he is no longer the only object of the derision.

Twenty minutes later, all of us—three wet men and one cold one—have stripped down to sweatshirts and jeans and thrown our wet gear and clothing in the back of the truck. I briefly imagine my coat and oilskins lying on bait and catfish and wonder what Cindy will think when she smells them in the garage. I cannot imagine the reaction I would get from the men if I asked to separate my gear from theirs because I didn't want it lying in frozen bait and fish. To suggest it might bring far more laughter and ribbing than tripping in the marsh could cause. I made the sacrifice deciding I would explain to Cindy later.

Driving back to Bob's we talk about various net fishing gear: gill nets are used in winter for rockfish and white perch; fyke nets are used for white perch, yellow perch, and catfish; float nets or drift nets are used for rockfish and sometimes perch; and pound nets are used for rockfish, menhaden, croakers, and trout. Other nets are used but these are the basic staples of the fishermen. They can catch perch twenty-four hours a day but find most success at night. Men will fish all night but it has been slow lately because of the water temperature.

"As the water gets colder the fish will develop a film over their eyes and it makes it hard for them to see the nets. That's when to go get 'em but we've had such a mild winter up to now that the Bay's kept the water temperature up. White perch, for example, will go to markets in Canada and California but they've fallen off here lately for a local market. Used to be that old folks would buy 'em and take 'em home to eat. This has had a big impact on the local market because young people don't like to do it," he says. "We've had a few good years now. 'Course we had our ups and downs but I think by and large we're seeing a major recovery in a lot of stocks. The secret for us is to keep the market up when the fisheries that have fallen off return good and strong."

When the trucks are backed up to Bob's shed door, he tells Amos that he wants the fish sorted and iced down for the deliveries he has to-morrow morning. Jimbo, Bob, and I then go into the house for a cup of coffee, a piece of pound cake, and a cigarette.

Twenty minutes later I say good-bye and leave the men at Bob's kitchen table talking about duck hunting and waiting for Amos to join them. I know their conversation will go on for another hour before Jimbo and Amos leave to drive to Dick's, a huge sporting goods store in Columbia, Maryland, an hour away, where they want to buy more gloves, shotgun shells, and other necessities to carry them through the next thirty days.

I remove my gear from the back of Bob's truck, now without bait and fish thanks to Amos. As I drive back to Fairhaven I'm confident that a day or so of drying in the garage will bring my wet clothes back to normal.

CHAPTER 4

Bay Fisheries

Winter sets in and watermen throughout the Bay region focus on oysters, fish, clams, maintenance and repair, legislation, and off-the-water winter jobs that range from leading hunting parties to being a merchant seaman. It is a good time to reflect on the state of the Bay from several perspectives.

The Chesapeake Bay is the largest estuary in the United States. It was mentioned by President Reagan in a State of the Union address, and it is often a major focus of discussion in national conferences and world-wide debates. Years ago it was rumored that Japanese dignitaries visiting "the great Bay" recognized that if "we had the Chesapeake Bay we could feed the world," a compliment to this famous estuary but, perhaps, an unrealistic one.

While it is true that resources from the Bay have made men wealthy beyond their imaginations and that towns like Crisfield, Maryland, have literally been built on foundations of oyster shells, those resources of decades past are being challenged and in many cases have been decimated or are crossing the threshold to extinction.

The driest growing season on record for Maryland occurred in 1999. Agricultural disasters were declared in most coastal states. Drought conditions actually began in the summer of 1998 when rainfall in the mid-Atlantic region and in the Northeast fell eight to eighteen inches below average. Temperatures were also at a record high, which added to the severe weather conditions.

While farmers were grappling with parched crops, many fish were dying in rivers and creeks. Abnormally high water temperatures and reduced water flow into freshwater tributaries of bays contributed to fish kills. Reduced water flow often results in higher water salinity, which is lethal to fish that are unable to escape. Low levels of dissolved oxygen in the water can also kill fish and usually results from high water temperatures, low freshwater flows, and related algae blooms.

An estimated 500,000 menhaden were lost in the summer along the lower Pocomoke River on Maryland's Eastern Shore due to low levels of dissolved oxygen in a tributary known as Bullbegger Creek. Bay scientists believe that the large school moved into the creek at night to feed or to avoid predation and became victims of the decreased levels of oxygen.

For Maryland's watermen, catch and therefore economics fluctuate year to year. The following list compares recent harvest statistics of some of the fifty fisheries the Department of Natural Resources tracks:

Channel catfish. In 1990, 1.8 million pounds were harvested with a value of $600,000. In 1998, 1.9 million pounds were harvested with a value of $827,000, with better market prices. Then in 1999 they dropped off to 940,000 pounds, a million pounds under the previous year, with a value of only $313,000.

Eel. In 1990, 155,000 pounds were harvested with a value of $267,000. In 1998, 281,000 pounds were harvested with a value of $412,000. Then in 1999 a disappointing 225,000 pounds were harvested with a value of only $360,000.

Spot. In 1990, 103,000 pounds were harvested with a value of $82,000. In 1998, 262,000 pounds were harvested with a value of $109,000. Then a small decline occurred in 1999 when 235,000 pounds were harvested with a value of $111,000.

Striped bass (rockfish). In the last year of the moratorium in 1990, 42,000 pounds were harvested (some released) with a value of $36,000. In 1998 with management plans in effect, 2.51 million pounds were

Eels squirm in a bushel basket waiting to be iced down and taken to a Jessup, Maryland, seafood packinghouse. They will be shipped to the Orient or sold to local customers who revel in the smoked or grilled meat. Courtesy Maryland Watermen's Association.

harvested with a value of $3.3 million. However, in 1999 only 1.16 million pounds were harvested at a value of $1.9 million.

White perch. In 1990, 826,000 pounds were harvested with a value of $384,000. In 1998, 1.4 million pounds were harvested with a value of $863,000. But in 1999, 1.4 million pounds were harvested with a value of only $647,000.

Blue crab. In 1990, 45 million pounds were harvested with a value of $19 million. In 1998 only 25 million pounds were harvested with a good market value of $25 million. And, while 26.5 million pounds were harvested in 1999, the value dropped to $22 million.

Oysters. In 1990, 2.8 million pounds were harvested with a value of $10.5 million. In 1998, 2.5 million pounds were harvested with a value of only $7.3 million. And in 1999 only 897,000 pounds were harvested with a devastating value of only $2.2 million.

Soft clams. In 1990, 2.1 million pounds were harvested with a value of $9 million. In 1998 a mere 227,000 pounds were harvested with a value of $1.5 million. The industry was barely standing in 1999 when 59,000 pounds were harvested with a value of $411,000.

One could cite many reasons for the rise and fall of harvests and markets. Men working the water will move from one species to another looking for the best market. Some of the examples mentioned above could show fluctuations due to natural causes or due to the men who work them, mixed, of course, with various new management plans and regulations that all tend to impact harvests and reports.

Striped bass had another above-average spawn in Maryland in 1999, the eighth straight year that rockfish have either rivaled or exceeded the long-term average since the five-year moratorium on catching the popular fish was lifted in 1991. Ironically, despite the recent glut of young fish, managers are worried that there is too much pressure on large, older fish in the population. Subsequently, they cut the coastwide harvest of large rockfish in 2000 by 14 percent.

This information is included in Maryland's juvenile index, the longest continually maintained record of fish statistics on the East Coast, and is considered an important indicator of future striped bass stocks because 90 percent of the coastal population spawns in the Bay. The index shows the average number of newly hatched fish, or young of that year, caught in seine nets during a summer monitoring program.

Unfortunately, the success rate in Virginia was not as high; it was the lowest since the moratorium of 1985. "We think it was the result of the drought," said Herb Austin, of the Virginia Institute of Marine Science. "We have good recruitment during the wet years but this spring was dry."

The balance of salt water and freshwater is critical to the resources of the Bay. High flows of freshwater into the rivers and streams during wet springs tend to drive salt water farther down the Bay, expanding the nursery habitat for young fish. A drought brings unusually high levels of salinity throughout much of the Chesapeake and its tributaries, dramatically impacting reproductive success.

While striped bass can live up to thirty years, few live past the age of seventeen in today's population. Large fish of spawning age, generally eight years or older, are needed to produce young and to maintain healthy stocks. The management program for striped bass is directed by the Atlantic States Marine Fisheries Commission, an interstate panel in which Larry Simns actively participates. Their goal is to have enough old fish in the population to ensure decent spawning success even if there is a series of years with poor environmental conditions.

According to fisheries scientists, improving the odds of spawning success requires not only a lot of older fish, but also fish of various ages. Older fish produce more eggs than younger fish, and different age fish spawn at slightly different times, increasing the chances that some portion of the population will meet up with good spawning conditions. These factors indicate that many of the fish hatched in 1993, which was a good spawning year, had not reached maturity in 1999, and were not large enough to reproduce or to be caught in the Atlantic coast fishery, where the minimum size is twenty-eight inches. Therefore the fishing pressure remains on the older fish, which are still in relatively short supply. Bill Goldsborough of the Chesapeake Bay Foundation sums it up by saying, "We're in a situation where we've got this bulge of younger fish that have not yet recruited to the coastal fishery in large numbers, so we have to be patient."

Patience is not necessarily a virtue when it comes to being a waterman trying to count on select resources to make a living. While watermen are known for their ingenuity and being able to jump from one fishery to another, they historically count on oysters, clams, crabs, and rockfish as their staples. Two of the four are scarce, which puts a lot of pressure on crabs and rockfish.

Few people think of menhaden when they think of fishing the Bay. It is not a fish caught recreationally; only select watermen fish for them, aside from the large factory trawlers working at the mouth of the Bay. These boats come from Reedville and Deltaville, Virginia, towns which were built on the menhaden industry. However, menhaden is a main source of food for rockfish. With growing populations of young striped bass, there is speculation that the population is outpacing one of its most important food supplies. Some fisheries managers think that the decline in rockfish in certain areas of the Bay is the result of a 79 percent drop in Atlantic Coast menhaden since 1991.

Menhaden are economically and ecologically important for Maryland and Virginia. They graze large amounts of algae out of the water as they grow and become a food source for larger fish. By weight, menhaden represent the largest commercial fishery in the Bay. And although not used for human consumption, the fish are used for their oil, for animal feed, and for other products.

Concerns over the menhaden population sparked the Atlantic States Marine Fisheries Commission to establish a peer review of the management plan in 1999 that called for establishing quotas on the commercial catch, taking into consideration the ecological role menhaden play in the Bay. This went even further as issues about the relationship of menhaden to striped bass and other predators served as a catalyst that caused scientists to rethink fish management. There are questions about predator-prey relationships among many species, including whether predation on blue crabs has increased in recent years. "It is a very complex food web," states Derek Orner, a biologist with the National Oceanic and Atmospheric Administration's (NOAA) Chesapeake Bay office. NOAA and the United States Environmental Protection Administration (EPA) have an office specifically dedicated to the Chesapeake. Nevertheless, in response to this concern, congressional intent in the budget of the fiscal year 2000 is to allocate $500,000 for NOAA's Bay office to begin to study multispecies interactions within the Bay system.

With growing stocks of striped bass, bluefish, weakfish, and other predators, there is a need for more prey. "The prey biomass needed to sustain those species needs to be higher than it was fifteen years ago," said Phil Jones, of the Maryland Department of Natural Resources. "We need to keep a close eye on this for the importance of menhaden ecologically."

A positive turn of events in 1999 was the first spawning run since the 1960s of American shad, a member of the herring family. Prized by recreational anglers for its fighting ability and once supporting the Bay's most valuable commercial fishery, shad had vanished from the Patuxent River and many of its tributaries in recent decades.

From an annual harvest in the Bay of 17.5 million pounds at the turn of the century, shad stocks fluctuated in a downward spiral to landings of approximately 500,000 pounds in 1999 from Virginia waters. Maryland, hit hardest by the decline, placed a complete moratorium on sport and commercial fishing in 1980 after the harvest fell to a devastating 25,000 pounds.

This fishery, like others—particularly oysters—has demonstrated the effects of civilized growth and development and natural impacts on its survival rate. Old-timers say that in many of the Bay tributaries shad boats were once so numerous that one could cross the river by stepping from boat to boat. Sadly, the American shad went the way of the once prevalent Atlantic sturgeon. Hope for anadromous (living mostly in salt water but spawning in freshwater) and estuarine fish in the Chesapeake lies in reversing the degraded water quality and habitat alterations that have resulted in decreased fish populations.

Recent stocking efforts have proven successful, as the shad population has begun to increase. According to Steve Minkkinen of the Maryland Department of Natural Resources, the Patuxent fish returned, almost like clockwork, five years after the first 14,000 fish were placed in the river.

For a fish that early settlers reported so abundant that they could catch them in frying pans, shad faced decimation because of overharvesting, pollution, and dam construction that blocked access to their spawning grounds. The recent resurgence is a promising sign for the Chesapeake Bay and for the team effort of watermen, fishery managers, and civilians alike. In 1999 various agencies stocked more than 27 million shad in more than half a dozen Bay tributaries. This is less than the 33 million in 1998 but above the average year stocking goal of 20–25 million.

In addition to hatcheries, major efforts have been made to build fish passages to reopen spawning grounds at dams such as the Conowingo on the Susquehanna River, which in some cases had closed rivers for nearly two centuries. Passages have now been successful in both Maryland and Virginia, and there is much excitement about the rebirth of this fishery. It's amazing to see efforts by humans—using so much time, energy, money, and dedication to help a resource—resulting in such positive change. There is a stretch of the Potomac River between Little Falls Dam and Great Falls, just northwest of Washington, D.C., which was a prime spawning ground for shad. With a fish passage being constructed in 1999, shad placed in the water above the Little Falls Dam will return, hopefully, when the dam is breached. "The shad have no genetic memory because the area had been closed since the 1950s," said Jim Cummins of the Interstate Commission on the Potomac River Basin. "That's another reason for the stocking effort, to imprint them" and have them move up the passage to the spawning grounds above the dam.

In May 2001, Jim Cummins called to tell me that the passages were completed in February 2000 but it was a difficult year for monitoring the shad population because accessibility to the ten-mile stretch of river through the rocks of the Falls was difficult. "We know some fish got up there but because it's so dangerous we can't get a fix on the numbers. We're going to try different nets and we will monitor the number of anglers and what they are catching. There are definitely striped bass, white perch, and shad." Cummins was optimistic that the passage was helping and would improve the species but was disappointed that he couldn't get accurate numbers in the first year.

The Bay and its tributaries continue to be studied by an abundance of scientists, educators, fisheries experts, and environmental organizations. They all point fingers at a variety of causes for a decline in water quality, harmful algae blooms, endangered fisheries, and so on. Natural occurrences such as the hot, dry summers of 1993 and 1994 can cause the water temperature to rise to a point that some fisheries, like soft clams, die off so much that there has not been a good spawn since. However, we can most frequently blame ourselves for the decline in fisheries. As the comic-strip character Pogo said in a parody of a famous quotation, "We have met the enemy and he is us."

People from all sides of Bay issues respect Pete Jensen, deputy director of fisheries for the Maryland Department of Natural Resources, for his expertise. "The overall health of the Bay is actually pretty good," he says. "We have a serious problem with a lack of dissolved oxygen which has affected spawning grounds in the headwaters of our tributaries because of a lack of flushing, a problem with nutrients causing algae blooms, and sedimentation. Folks in the general population will say, 'We get the grasses back and that will change the state of the Bay.' While that may be true to a certain extent, grasses are merely an indicator of the Bay's health. The real problem is societal," he states.

"From my perspective the largest threats to fisheries and the watermen remain perceptions about water quality and overharvesting, respectively. I can see healthy spawning areas where populations can beat a serious water quality problem, but you can't fish those areas," Jensen says.

I ask how we can fix some of the problems that face us now. He emphasizes population studies that indicate the trend is for more people to move within fifty miles of the coast, causing an increase in development, which will result in negative impacts on the resources. "Development is influenced by money and politics, or maybe it's vice versa, but regardless, you can't change it because of the economic impact it has on local and state communities. You must channel it so that proper decisions are made and result in less impact on the environment," he said. "A real problem for us is identifying fishery mortality and defining it in terms of water quality. We can measure mortality via overharvesting but not in terms of water quality, and it's frustrating," Jensen concluded.

For the watermen, one of the most disturbing occurrences of 2001 was the announcement that Jensen—the state's major proponent of reason and fairness among fisheries management, watermen, and recreational anglers—resigned in April without explanation. Rumors of "forced" retirement were quickly followed by "no comment" from the department. Pete understands the watermen perhaps better than anyone in state government, and his resignation is a great loss.

CHAPTER 5

CBF: Controversial Protectionist

The Chesapeake Bay Foundation, a nonprofit watchdog group with considerable financial and political influence, has caused concern among the watermen for years. Its annual State of the Bay report has wide and significant distribution and is accepted by thousands of Bay activists who frequently criticize the watermen for overfishing, damaging the environment, and so on.

While few can deny that the Chesapeake Bay Foundation is an excellent organization performing its role with expertise, some bureaucrats as well as watermen claim the foundation puts a negative spin on issues for the benefit of member growth and support when there are positive developments that could be touted as well. One could go so far as to say it is a love/hate relationship the watermen have for the foundation. Watermen appreciate the efforts of CBF to restore oysters and protect the crab population, as well as their use of staff and boats to undertake studies and to help with ecological monitoring, pollution reduction, habitat protection, and more. The watermen have been very active and supportive of CBF in the area of environmental education. The foundation serves about 40,000 students—adults and children—each year through their on-the-water education programs.

I have supported the foundation for a long time and am a strong believer in the efforts of its staff. For many years it was the only organization available if you wanted to be a citizen advocate and wanted to effect changes to alter the declining quality of the ecosystem. CBF has taken on issues ranging from restrictions against pollution by Bethlehem Steel to regulations on local fishery practices. Like any successful organization determined to make a difference, it cannot avoid controversy.

For example, several years ago when the foundation pushed for regulations on harvests, particularly on crabs and oysters, there were signs all over tiny Tangier Island, Virginia, not to support CBF. On Smith Island, Maryland, up the Bay from Tangier, a shed at the CBF education center was mysteriously burned and its boat vandalized.

These isolated islands have roots going back to the 1600s. Residents of Tangier still speak in a broken Elizabethan dialect. They have survived for centuries on the harvests of the Chesapeake, and it seemed to them that the Chesapeake Bay Foundation was getting in their way. That changed, at least temporarily, when Susan Drake, a doctoral student from the University of Wisconsin, moved to the island in 1997. Ironically, Drake was not only an outsider and environmentalist like so many before her, she was also a woman entering a man's world. But Drake had more going for her than her predecessors did; she was an evangelical Christian, and for the people of Tangier Island, religion is paramount.

One of Drake's earliest supporters was Carlene Shores, who said of Drake, "She was much more easily accepted because she spoke in religious terms. She wanted us to see that it was our duty as Christians to protect the environment and look after God's creation."

Drake wanted to understand the islanders' hostility toward the Chesapeake Bay Foundation. She quickly realized that there were a number of factors involved—for instance, the conflicts of blue collar versus white collar and educated versus those less educated—but it all boiled down to their fear of losing their way of life. She started talking to members of the island's two congregations and began organizing committees as a way of getting a firmer grasp on the community's concerns.

By February 1998, after Drake had delivered a formal talk at a joint service of Swain Memorial United Methodist Church and the nondenominational New Testament Church, fifty-eight Tangier watermen agreed to sign a Stewardship Covenant—a promise to themselves, to other watermen, and to God to "keep all laws, particularly fishery, boat, and pollution laws, in order to protect Tangier's cultural heritage and ensure a future for the next generation." This, it seemed to many, also fell in line with the intentions of CBF. Curious to many in the Bay community was that each man who signed the covenant was given a red ribbon, signifying Christ's blood, to fly on his boat's antenna.

Meanwhile, as momentum continued, the women of the island formed an auxiliary organization called FAIITH (Families Actively Involved in Improving Tangier's Heritage). Carlene Shores was elected chairman. FAIITH established three immediate goals: first, to collect regulatory and scientific information on fisheries and disseminate it to watermen's families; second, to ensure waterman representation by attending key meetings of the Virginia Marine Resources Commission; and third, to collaborate with government and advocacy groups to find solutions to problems threatening watermen's heritage. They looked upon groups like CBF as potential advocates rather than adversaries.

Continuing to tightly weave biblical beliefs with environmental responsiveness, the citizens of Tangier were committed a year later to improve their home, and in the spring of 1999 more than seventy people turned out for an unprecedented islandwide cleanup effort. At the time, I recalled trying to help spearhead a cleanup campaign a decade earlier on Tilghman Island, inspired by teachers and students at the local elementary school and also involving St. Michaels Elementary. My efforts were meaningless as the cleanup was quickly and quietly canceled before the first McDonald's wrapper or battery could be removed from Knapps Narrows. Not a word was ever spoken; it just didn't happen.

Residents of Tangier Island tried to avoid publicity about their community and their cleanup efforts. They wanted to keep their community's affairs private, and many resisted when Shores, Drake, and others began an environmental crusade that brought unwanted attention to the island. A new sign appeared on the island reading, "Due to the Chesapeake Bay Foundation Tangier's Heritage is Deeply in Trouble . . . Do Not Support . . . Thanks." Waterman Ken Pruitt, speaking with the *Bay Journal,* said pollution was not a serious problem. The crab population simply fluctuates in cycles, he said, and all anyone can do is "take the bitter with the sweet."

The wife of another waterman, who did not want to be identified, said she obeyed the pollution laws and there was no need for Drake to come to the island and "shove the covenant down our. . . ." She also felt the red ribbons divided the island residents.

A similar difference of opinion is evidenced by the islanders' reaction to two other issues—oyster restoration efforts and crab management initiatives by CBF and others. Perspectives and loyalties can easily become clouded on small island communities, particularly when the livelihood of the residents often depends on decisions made on the mainland. Like many in the waterman community, island residents are desperately concerned about the oyster population and want the replenishment activities to continue. For too many years, productive oyster beds were destroyed by MSX and Dermo, cutting off needed income. The CBF plan (backed by state funding) to establish brood stock sanctuaries and rehabilitate harvest areas shows promise that the future may be better for this fishery, giving islanders hope for the restoration of Tangier Sound's oyster population.

However, the residents' view of crab management is not as positive, and this fishery is a huge source of revenue for watermen. Some people understand and agree with the foundation's approach for managing the crab population, while others see it simply as a means of restricting their way of life. If a CBF staff member discusses crab management in broad terms, watermen may focus on only one comment and consider it relative to their own backyards. Taken out of context, a statement can prompt unfair criticism of the foundation.

Unfortunately, residents of the islands, like many watermen around the Bay, sometimes don't see the forest for the trees. They see their local situation in great detail, but they might see a very different picture if they looked at oysters, crabs, or other species Baywide. This, to an extent, is where the foundation and even the Maryland Watermen's Association may be misunderstood. Both must look locally but direct a greater concentration on the big picture.

While Tangier resident Rudy Shores and some other watermen see CBF as a possible new backer, they see the Virginia Marine Resources Commission as a threat. "We've survived periods of low catches and low prices for a hundred years," says Shores, "but the commission can easily regulate us out of business." CBF influences regulation, but the commission, like the Department of Natural Resources in Maryland, makes the regulations. Watermen see the difference and by the end of 1999 some islanders were cautiously optimistic that environmental advocates wanted to work with them.

Not everyone shared the optismism. In the January-February 2000 edition of the *Bay Journal* Carlene McMann Shores wrote a piece called "Tangier Watermen Feel Betrayed by CBF." Shores's testament on betrayal and disappointment focused on CBF efforts "to push for a continuation on the freeze, transfer and upgrade of licenses. They also promote a reduction in both the peeler and hard crab pot. We, and many water communities, have young school boys, young men and even older married men with families to support . . . who cannot work on the water because of this freeze."

Shores continues, "By mistake, we slept with the enemy. We were deceived and our trust was abused. If we don't make a stand, we will lose what we treasure and love the most.

"As Christians, we will continue to do these things [to preserve our heritage]. We often feel as the writer of Corinthians must have when he said, 'We are hard-pressed on every side, but not crushed; perplexed but not in despair; persecuted but not abandoned, struck down but not destroyed,' " she concluded. I was moved by this description that so adequately describes what Bay watermen have felt for as long as I can remember.

In truth, while CBF did support the temporary freeze on licenses, it always supported an exception from the freeze and limited entry for

Tangier watermen, an important and often-unpublicized point. When the Tangier exemption was shot down by the Virginia Marine Resources Commission, CBF continued to support it. "It appears that what Virginia needs is an orderly process for limited entry that will allow young men and women to follow the water and this hasn't happened yet," said Bill Goldsborough, CBF's senior fishery scientist.

When I ran into Goldsborough at the Commercial Fishermen's Expo in the waning days of January 2000, I was most interested to hear his side of the story. Bill works closely with watermen throughout the Bay and has spent significant time on the islands.

The foundation has been a player in island society for over two decades. Goldsborough opened the first CBF Environmental Education Center on Smith Island and lived there for two years starting in 1978. Today staff live on both Smith and Tangier Islands; dozens of residents have been employed by the foundation; and the foundation's educational field trips, which have involved thousands of children and adults, fuel the local economy. In addition, CBF helped design the Smith Island crab co-op and assisted with creating its business development plan. The foundation has put more than $1 million into capital investment at its environmental education centers on the islands, a significant price tag for an effort to work with the communities and their residents in improving the environment.

"One of the reasons that we, the foundation, end up creating controversy is that we must look at the big, long-term picture and sometimes watermen don't do that," Goldsborough states. "The Bay cannot produce the valuable biomass it did years ago. The food web has shifted today with an overabundance of algae, which shades the submerged aquatic vegetation (SAV) causing tremendous damage. This situation must be reversed and we're making progress with replenishing grasses one step at a time. I think we're on the cusp of a huge change for oysters and SAVs. The oysters bottomed out in 1994 due to disease and other factors, but I think we're turning a corner. Scientists, managers, and watermen hold new optimism. The new initiative, a goal of the Chesapeake Bay Agreement to increase oysters by tenfold in ten years, is terrific. We need to put resource dollars into rebuilding habitat and seeding programs. Twenty-five million dollars over ten years for Maryland, and the same for Virginia, with federal matching dollars creates a lot of reason to be optimistic," he says.

Goldsborough explains that there is similar hope and a similar massive public involvement effort in place for the grasses. "While we still have problems in Tangier Sound where 60 percent of the SAVs have been lost in the last seven years, patterns are changing, and now we're seeing grass come back where there are people, like the heavily populated Severn River. The Tangier problem, like others on the Eastern Shore, could be attributed to agricultural runoff way upriver."

When I ask him about the crab population and the conflicts with watermen over this valuable resource, he says, "We have a million questions about them [crabs] and when we think we know something they fool us completely. But a ten-year winter dredge survey on a thousand sites around the Bay, which was a collaboration between watermen and scientists, gives us a snapshot of the crab population and there are very good predictions for the future. Much better than the *Farmers' Almanac!* Truth is, if crabs weren't so resilient they would have collapsed a long time ago. The watermen were forced to rely more heavily on the crab resource because of the poor oyster and clam harvests over the past decade or so. And even though the large crabs aren't what they used to be, I think we're turning another corner because we're getting a better picture to substantiate it.

"Limited entry, we believe, is essential to all species, and watermen must make a good living. Limited entry, in one form or another, is the only way to do it. I think you'll see this management tool increase in popularity all up and down the coast. When the stock gets smaller and smaller, limited entry adds stability to the efforts of harvesting and making money. The Bay Commission conducted a valuable two-year study, a first-time effort, under the Bi-State Blue Crab Advisory Committee. Because it's all one crab population, both states, Maryland and Virginia, had to be involved. In 1999 the committee started tabulating information to see if there are better ways to manage crabs. I support that approach because the goal is to see watermen make a good living with less effort, to be able to reduce the number of pots and still be successful.

"When I look at these efforts and what we're trying to do, I do think of FAIITH and the people on Tangier. That's a great stewardship effort based on scripture. We have a good relationship with the folks on Smith Island, and Janice Marshall [a Smith Island resident] is on the CBF board. Chuck Marsh is president of the Smith Island watermen's association and a good man. Anti-CBF feelings seem to come and go with the issues. If they like what we're doing then they are fine; if not it reverses back to 'get off my turf,' " he states.

"Look, we're dealing regionwide and there are differences between the Maryland watermen and those in Virginia, and those in Baltimore County are as different as night and day sometimes from those men from Somerset County. Larry Simns's legacy to Maryland watermen is that their association is unified and viable, but in Virginia they never could get a state association off the ground and the local counties fight amongst each other. We're somehow in the middle of all this more often than I would like.

"The watermen want to have their cake and to eat it too. They want high reproduction of all species and high catch limits. The foundation has the luxury of taking the long-term view because we don't depend on the resources to make a living. The watermen don't have that luxury and, consequently, we're often at odds," Goldsborough concluded.

It will no doubt be years before we have solutions to many of the problems related to the restoration of the Chesapeake's ecosystem, but what is certain to me today is how unfortunate it is that watermen don't partner more often with scientists, and scientists don't partner more with watermen. Two leaders in the community—Larry Simns and Bill Goldsborough—agree, and that is a big step. When more people set aside their prejudices and stop finger-pointing, some proactive things may start to happen. Simns and Goldsborough will then be the first of many catalysts for positive change.

CHAPTER 6

Watermen Turned Politicians

In response to the increasing amount of legislation affecting Bay fisheries, watermen began leaving one culture for another, no longer working the water but instead working as those who impact the watermen's way of life. Some men became so active and intrigued by the political arena that they traded their oilskins for coats and ties forever.

FISHERMEN'S EXPO AND GEORGE O'DONNELL

Before crossing the Chesapeake Bay Bridge just after sunrise on January 26, I look out over the Severn River, sit back, and relax. I anticipate an uneventful two-and-a-half-hour ride to Ocean City, Maryland, for the East Coast Commercial Fishermen's and Aquaculture Exposition which is sponsored and run by the Maryland Watermen's Association. I have a booth at the expo for a weekend book signing, but my goal this year is to spend as much time as possible interviewing a few key people for this book and making arrangements with watermen to go out with them over the next month or so.

At the Bay Bridge I pay my $2.50 toll and let an eighteen-wheeler slowly edge his way in front of me. I position myself next to the outer guardrail as the incline increases across the shoreline and over the Bay. I subconsciously employ my habit of looking through the silver-painted guardrails that reflect the winter's morning sun, watching for clammers or oystermen working the beds off the beach south of Sandy Point State Park. The Bay is vacant this morning, totally void of workboats as ice covers most of the water. The main channel is open and I watch two freighters from foreign ports make their way slowly up the Bay to Baltimore Harbor. Where there are holes in the ice, sunlight sparkles off clear, cold water, now below thirty-two degrees, and surface windchills dip to ten degrees below zero. Seabirds rest at the openings and I imagine scenes from Discovery Channel programs in which a polar bear pokes his head up through the ice. While Bay folks have been surprised to see a submarine, a piranha, or a stray Florida manatee break the surface, a polar bear is most likely out of the question!

The expo is held each January in the Ocean City Convention Center and draws commercial watermen, charter captains, offshore fishermen, and curious people from the mid-Atantic region to a commercial boat show and accessory arena. Manufacturers of gear, engines, lifesaving equipment, workboats from as far away as Nova Scotia, and recreational boats for fishing or pleasure share exhibit space with bait companies; clothing companies; insurance companies; makers of Bay crafts, books, decoys, and more. These vendors offer their products or services to men and women who will spend thousands of dollars during the three-day event. Betty Duty of the Maryland Watermen's Association is the matriarch, ruling over dozens of volunteers who helped set up the show, and managing everything from taking admission money to selling merchandise from the Maryland Watermen's Association booth.

As I pull into the parking lot the first thing I notice is the contrast of transportation that I see at numerous recreational boat shows I attend each year. Rather than BMWs, Volvos, and minivans, the convention center's parking lot is full of pickup trucks, bathed in dry salt from the recent snowstorms and many with refrigerated cabs on the beds.

For many of the attendees this is the one event of the year when they gather with wives, girlfriends, and often children to socialize with others from around the region. For most, it will be the only time they catch up with each other until the expo next year.

Expo attendees visit the Oyster Recovery Partnership booth to gather information on recovery plans, the state of oyster replenishment, and plans for expanding the program in the future. Courtesy Maryland Watermen's Association.

In years past, the main event for the attendees was the dinner dance held at the Sheraton Hotel. While that was suspended a couple of years ago due to lack of attendance and the cost of the event, the attendees still pack the Sheraton's main ballroom for a reception on Friday night to devour Bay seafood and vegetables from three or four buffet tables. They will be decked-out—not quite as they would be for a dinner dance, but acceptably so for an informal reception by anyone's standards. Here, packed in like sardines, they eat, drink, and look for old friends. I attend the reception every year to mingle and eat. It's my time to catch up with my childhood friend, Terry Chones, a regional sales manager for Suzuki outboards who lives in Annapolis and attends the expo every year. While we see each other several times during the year, this allows us three days of quality time to talk and to enjoy ourselves in the company of men and women who represent the salt of the earth to us. We'll have some fun!

As the crowd continues to grow throughout Friday afternoon, men in work boots, hunting boots, rubber boots, or docksiders stroll the aisles like kids in a candy store. The cash flows as if they were in a casino, and they walk out, making trip after trip to their trucks, with plastic tubs, padded deck covering, paint, rope, foul weather gear, crab pot supplies, and more. The purchasing will continue uninterrupted until the show's first day winds down and the men ride up Coastal Highway, a few blocks from the ocean, to the Sheraton to pick up their partners to get ready for the reception.

Attendance on Saturday is strong, which is fortunate for the Watermen's Association as this is its major carnival fundraiser. Raffle tickets are on sale for a chance to win a new, fully loaded, Ford pickup truck. The tickets are $100 each, and only 400 are sold. I buy a ticket every year confident that it finally will be my turn to drive away in a new truck. The tickets sell quickly.

I run into Kenny and Karen Keen and we chat for a few minutes. They have left their daughter in Chesapeake Beach so they have a weekend away. By Sunday morning Karen will be regretting an overindulgence in dancing and beer and Kenny will be laughing at her as she ends her weekend moving more slowly and less enthusiastically than when she started. She will not be alone by any stretch of the imagination.

I'm sharing a booth with lawyer-turned-author Tim Junkin from Potomac, Maryland. Junkin spent time on the Eastern Shore growing up and went to high school in Easton. His novel, *The Waterman*, has been out since September 1999 and is doing very well. We will take turns covering the booth to sell each other's books during the three-day affair. This works out well for me because I want to spend time talking to the men.

Bob Slaff, my close personal friend of many years and a reporter for several publications on the Bay, approaches our table. He wants to do an interview with Junkin and to spend time with me. Bob and his wife, Ester, are in their seventies and have been married for over fifty years. They

Watermen take a social break during the annual East Coast Fishermen's and Aquaculture Exposition held in Ocean City, Maryland, each January. The event is sponsored by the Maryland Watermen's Association.

still act like teenagers, have twice the energy I do, love to go out on the Bay in one of their three boats, and attend the expo every year to support the watermen. Bob loves telling the watermen and others that he is a seventy-six-year-old, Jewish, redneck charter captain with a handlebar mustache—and the only one on the Bay. I think he's right.

George O'Donnell, a waterman and oyster diver for twenty-two years before entering politics, stops by my booth on his way to see locally famous artist and charter captain Tilghman Hemsley. Hemsley is highly respected for his art and his personality by virtually everyone who knows him. He has designed a sculpture of a working waterman that will be placed on Kent Island. O'Donnell is a big supporter of Hemsley and will help sell raffle tickets to raise money for the sculpture.

George O'Donnell is a large barrel of a man. He is soft spoken, philosophical, and long-winded. When he begins a story, you must be prepared to hear the end of it. He ran a successful campaign for county commissioner of Queen Anne's County because he "wanted to help watermen fight for their way of life."

"I didn't like the idea of getting off the water. I've tried it before and it's a hard row to hoe because the water is a calling, a way of life that once it bites you, it owns you," he says.

"When I first ran for county commissioner I lost by eighty-two votes. I realized then that I couldn't work the water and be a full-time politician. I made the hard decision to quit the water and got a land job as a food service rep. I was nervous, you better believe, never having touched a computer. Well, it wasn't long before I realized my gift was in talking. I quit the rep business because of traveling and major issues going on in the county. I'm in the security systems business now and more comfortable with my place now.

"When I started diving there were maybe a hundred divers on the Bay and about five thousand shaft tongers. We were a new gear type and

Since turning in his oyster diving gear for a coat and tie as a Queen Anne's County Commissioner, George O'Donnell has fought hard for the Bay and for the watermen. He was one of the first outspoken politicians to jump into the fray concerning the use of Site 104 as a site for dredge spoil from the barge canal into Baltimore Harbor. Courtesy Camille Dove.

the shaft tongers resented us invading their territory. It's funny but on the water you can have clammers fighting oystermen, one gear type fighting another, and all of them fighting the forces that are trying to put them out of business. Unbelievable, sometimes.

"This whole thing with the shaft tongers ended up with us having zones where we could dive and not invade on the shaft tongers. We worked it out, you see. Then we had a problem where the state wanted divers to have a four-inch cull size when everybody else had three inches. We challenged that in court and won the case based on the law being unconstitutional. DNR had to change the law in the mideighties.

"There's a way around everything when you're trying to scratch a living off the water. When we still had the four-inch cull limit but the tongers could keep three-inch oysters, we'd get on the edge of a bar in three feet of water and shove the three-inch oysters off the cullin' board and when they were overboard we'd pick up a pair of hand tongs and bring 'em back into the boat as legal oysters. All this wrangling got me involved in the legislature with Larry [Simns] and the political process. I learned early from Larry that if you present good arguments they will make good decisions. I was turned on to making changes. You know there's not a soul in the Maryland legislature who is or was a waterman. This may drive me to run."

O'Donnell is lining up his political agenda. He will soon be president of the Maryland Association of Counties and, he says, "as the leader of the 'club' I'll be able to have some influence.

"We have to look way down the road at the big picture if we're going to make a difference in the Bay's health and her resources," he continues. "Now I'm concerned about the financial stability of the Watermen's Association and what will happen when Larry and Betty [Duty] retire. The walls are closing in on commercial watermen and it's not 'cause of a lack of resources, in my opinion. The CCA [Coastal Conservation Association] and others don't want any commercial fishing because it interferes with their recreational fishing. They have a lot of money and are organizing recreational fishermen into voters. No, in my opinion, the resources aren't the problem. Ecologically the Bay is coming back. There's

A diver goes overboard for the biggest and best oysters in the Bay. The water is cold and murky and the man must feel with his hands for the pick of the litter. It is dangerous work; the diver relies on a thin tube of air that can easily get tangled and create a life-threatening situation. During his twenty-two-year career as a waterman, Queen Anne's County Commissioner George O'Donnell was an oyster diver. Courtesy Maryland Watermen's Association.

more rockfish out there than you can shake a stick at and there's more white perch than ever before in my lifetime. Oysters are coming back in the upper Bay. No, the problem isn't that the Bay is 'dying' or watermen are overharvesting. How can they overharvest when there are so many regulations on them and their catch? Regulations and economics dictate how much crop they will take to keep their interest. They will get in and out of one product to go to another if they might make more money. It's the only way to survive.

"Still, overharvesting or reduced stocks are the last things that will kill a waterman. There will always be a seafood industry, in my opinion, but the real pressures are coming from recreational boaters and fishermen, environmentalists, and residents forcing regulations, which will shut down the economics that support the lifestyle. Watermen need to be more unified and show less individual greed on specific issues, even in the legislature," he continues.

"And, I'll tell you, undereducated legislators are a big problem. Larry does an unbelievable job at educating the key ones but every time you turn around there's a new face or restructuring of our key committees. Why, a woman asked me in a hearing, how I can dive for oysters when they're buried in the mud. I told her 'Ma'am, that's a clam!' And here is a person voting on an issue dictating my livelihood! Nope, I'll say it again. There's not one legislator that is or has ever been a full-time waterman and this is the kind of stuff that may make me go for it," he concludes. We both know it's time to get back to our booths. We promise to talk again. He leaves and I am as confident about George O'Donnell's continuing to climb the political ladder as I have been about anything in my life. For the sake of the watermen I pray I'm right.

The weekend proceeds in a similar manner and I'm having a great time, although I'm not selling many of my books. I've talked to many old friends, lined up some boat trips and interviews, and bought a new pair of boots that hopefully will keep my feet warm when I go out on the water.

Betty and Larry decide to close the show early on Sunday because of sleet and snow moving into the Delmarva Peninsula. Many of the exhibitors and attendees have already left for Delaware, Virginia, North Carolina, and other parts of Maryland to beat the weather. I leave as soon as I lose the raffle drawing for the truck and have an easy drive until I hit Easton where sleet and ice are beginning to cover the road. It takes me an extra hour to get back to Fairhaven in the bad weather, but it was worth the trip.

BUDDY HARRISON

Buddy "Little Bud" Harrison comes from a long line of watermen. His family is the hub of the seafood industry on Tilghman Island where they own Harrison's Oyster House (wholesale fish and oysters), Harrison's Chesapeake House (hotel, bar, and restaurant), a nice fishing charter fleet, and other businesses, including Bud's wife Leslie's Island Treasure Gift Shop. At forty-two, Bud is in his first term on the Talbot County Council. He is a highly respected charter captain, a paramedic, and former chief of the Tilghman Island Volunteer Fire Department; he's thinking of running for the state legislature in six years. His father, Buddy Harrison, whom I call the J. R. Ewing of Tilghman Island, with his diamond ring and custom cowboy boots, took the Harrison seafood business legacy to a political and economic plane well above what anyone would have imagined on the tiny island. Little Bud learned a lot from his father, and his workaholic waterman ethic goes into high gear frequently.

The day I caught up with him he had just completed a health inspection of the restaurant and was preparing for a fishing party due to arrive at lunchtime. I considered my thirty minutes stolen from his schedule a valuable commodity.

The Harrison name is widely known in the Tilghman area, with varying degrees of respect, but Bud Harrison had to run his race for the council countywide. "There are no district races here. I had to spread out and make myself known. I went everywhere from black churches to small towns but I was determined to give us [the Tilghman area] a voice on the council that decides our fate [because Tilghman is an unincorporated town, the county council determines its laws and regulations]. It was time that small business and the watermen had a voice because for years the council was made up of retired people or people without a small business background," he says. Bud Harrison was elected by a substantial margin in 1998.

"The biggest changes we have are the outsiders coming in. They've changed things around here. With development gone mad on the island, local people who have to work for a living every day can't really afford to live here. And, in the county, local watermen have lost their slips to big recreational boats. It's not anybody's fault, certainly not the recreational boating industry's. I know they don't have any choice. But these people will come in and say, 'We don't want your smell, your mess, or your noise, but we sure do want your seafood!' Too many of our people can't live here and can't keep their boats here. That's why I wanted to get a voice. The county has to do something to help, especially with the boats," he explains.

"I know the seafood business hasn't been what it was. Now, last year was a good year for oysters and this year looks a little better. Not the thousand-bushel-a-day my father bought in the seventies but it's okay. He taught me that in business you have to adapt to change and with that decline in oysters over the past ten or fifteen years I've shifted over to buying perch and rockfish. In fact, probably 50 percent of my business in the oyster house was fish, not oysters.

"The development issues are the toughest thing I have to deal with. Half the people have money or are retired so they don't want any more development 'cause they got theirs. And half, the working people, want development so they have a place to live and shop. I'm not sayin' we need a big box [of a building] built over here but a Kmart or the like would be welcomed by a lot of people."

Bud Harrison has learned to focus on his responsibilities to his heritage, his businesses, his island town, and his constituency. The Harrisons have long been known to lead fundraising drives, most recently to help Captain Wade Murphy with the salvage of his sunken skipjack. Bud does this in a quiet and yet forceful way. He knows and accepts the challenges associated with the role of a politician to make sure

Bushel baskets overflowing with oysters are unloaded at Harrison's Oyster Company on Tilghman Island. A waterman can consider it a successful day if the quota was reached and the market price was adequate. Courtesy Maryland Watermen's Association.

that the watermen are counted when decisions are made. He would probably do well in the legislature but if he decides to follow that path he will have to leave the local issues to someone else, and that is where and when he will have to make the hard choice, because who will follow in his footsteps for the island area and the watermen?

RONNIE FITHIAN

It had been several years since I last spoke with Ronnie Fithian, the waterman I think could pass for Nick Nolte's twin. I worked seed oystering with him and was impressed with his desire to be successful and with his love of the water. Fithian spent years as a clammer, fisherman, and oysterman when making $2–3,000 a week was the rule rather than the exception. In the mideighties when the market was good, he could gross $100,000 a year. Then things changed.

"In the early nineties I decided the water business was going down-hill to a degree that I didn't want to be on the water when it ended. I ran for the Kent County Commissioners in 1994 and won. There are only three of us, and I'm in my second term. Now I'm serving as the president of the board. I sold my boat and stayed workin' but then in 1996 the job of Rock Hall Town Manager came open and I applied. I serve at the pleasure of the mayor and town council, and I've been here ever since," he said.

"Rock Hall was once a waterman's town. Why, we had forty-six clam boats tied up in the harbor year 'round. Now there's neither a clam boat to be found here. The town went through transitions when the commercial fishing came to a halt. Every day I see men leaving the wa-ter. Ten years ago I sold clams to fifteen different shucking houses in New England. Now they've all but shut down like the crab and oyster packinghouses on the Bay. Why, we used to haul so many oysters down to Virginia it wasn't funny. Those shuckin' houses are all gone except for folks like Bevins. Only the big companies made it with clam shucking. People clammin' since it came into the Bay are getting out. It's all they done. Now what?" he asks rhetorically.

"We're seeing the changes like other small water towns. Outsiders come in to visit and before you know it they decide to stay. Once that started happening commercial docks were replaced by recreational ma-rinas, and property values went through the roof where a local boy couldn't afford to buy his parents' home!

"Marinas replaced the seafood business for the most part, but I'll tell you, it was a blessing. When commercial fishing dropped off and men were looking for work they found it in recreational boatyards and mari-nas. If it weren't for those marinas and the big recreational boats comin' in here, there'd be tumbleweeds rollin' down the main street of Rock Hall."

One of the major proactive initiatives that Larry Simns set in mo-tion for the Maryland Watermen's Association over two decades ago was to establish a long-term and fair relationship with the recreational boating industry. While there are times the recreational community dis-

Captain Ronnie Fithian was a lifelong waterman until he took the job as town manager for Rock Hall, Maryland, a "waterman's town" with a history steeped in the tradition of commercial fishing. Now he discusses zoning and water issues and looks out for a large constituent base—the watermen. Courtesy Ronnie Fithian.

agrees with the commercial side of the aisle, they always agree to disagree and then move on in as close a partnership as possible. Many times over the past twenty years I have relied on Simns and the watermen for guidance and help in the regulatory arena and they have never let me down. Bob Slaff, patriarch of the Maryland marine industry, former president of the Marine Trades Association of Maryland and now a freelance reporter, often told me, "We must work with Larry. They deserve their part of the Bay and to coexist is the only solution to our long-range concerns. We can't afford to fight them and we have no business fighting anyway. We need Larry perhaps more than he needs us."

Fithian agrees and tells me, "Ten to fifteen years ago we had a little adversarial relationship with the marina and boat salespeople. They were changing our town. We were antagonistic to 'em. Now, by God, I'd have to say they saved this town. Bringing in tourists and money created new jobs and we're doin' all right, even if we can't make it altogether on the water anymore."

Fithian has his own thoughts on what went wrong with commercial fishing and, like many others, he mainly blames the government. He explains that a large population of fish in the Bay does not really help the watermen as people would believe. "The way the state has set up fishing quotas and seasons doesn't help us at all. There's a whole basket of reasons why you don't catch fish even if they're plentiful: temperature, water clarity, wind, migration, and so on. Three-quarters of your fishing days, you won't catch nary a fish. We should be allowed to catch all we can on any day to make up for the days we don't catch anything. The state messed up our livelihood. They are constantly overcorrecting the problem instead of controlling it. When the rockfish were scarce, they imposed a moratorium. Then when the fish came back they became so conservative with quotas and times to fish that they messed it up for everyone. They shouldn't have turned us loose, but they could have relaxed a bit.

"They did the same with goose hunting, which is big business on the Eastern Shore. When the geese came back the state said, 'Take three.' The commercial guides said, 'No, we just want two.' The state insisted on three and now there's been a moratorium for years because the population went down. There ain't no season at all now. You can't figure the state out, so I said to hell with it. And, I'll tell ya, mark my word now. The same thing will happen to the crabbers. This talk about crab sanctuaries and so forth will lead to the end of that fishery. See, the crabbers saw what happened to the fishermen and they won't work with the state, I don't believe. They know they won't ever get back what they give up, even if it's with good intentions in the name of the fishery." There is a part of me that, regrettably, thinks this will become a truth.

"I think about goin' back on the water from time to time," Fithian says, "but there's so many laws now you couldn't break 'em all if you had to. I'd have to take a class to become a law-abiding waterman. Ain't that somethin'?

"Here's how screwed up it is: My fiancée is an oyster buyer and here we are on September 18, 2000, and she doesn't know when oyster season officially starts! I believe the rumors are true that it will be September 25 at ten bushel a day to make up for last year's short season, but nobody knows for sure. If they're changing a season they need to tell everybody so they can be prepared. There's trucks, booms, boats, winches, ice, everything to get ready."

When we shifted the conversation to the environmental condition of the Bay, Ronnie asked me if I had another couple of days to talk. "I read all that information on the state of the Bay. We've got a lot of zoning and development issues up here that tie into it. And I don't care what they report or what you read; it ain't agriculture and it ain't industry, and it sure as hell ain't the watermen that's doin' the Bay in. It's people—people and politics. DNR can't control the resources. They don't know what they're doin' and so, when a man can catch more fish because there's a lot out there, DNR isn't goin' to let them because that wouldn't be politically correct. Hell, they should let the recreational fishermen catch more, too.

"The key is to have grasses to have a healthy bay. The grasses are the key. Find out what's killin' it and we'll solve all the major problems. They are the nucleus for the survival of so many species. Grasses used to be so thick they acted as a buffer in the harbor against a sea. Now there's none. Now they can blame the clammers but let me tell you the truth about that. There used to be good grass on the Susquehanna Flats and there's never been a clammer one up there or up any river so you can't blame 160 damn clammers for takin' the grasses. Put the blame where it belongs," he says emphatically.

I recall growing up in St. Mary's County when grasses were so thick that we had to row our motorboat from the pier out to the channel on the Potomac because it was impossible to start the outboard without grasses fouling up the prop. Grasses were so thick that our neighbor Mrs. Bannigan would walk along the shore and tow behind her an inflated inner tube supporting a bushel basket that she would fill with soft crabs caught with a dip net. Oysters were so thick on the shore that my father would direct us to go overboard and toss him up a few while he sat on the pier with an oyster knife and shucked them where he sat. The clammers didn't destroy the grasses; we did as we developed upriver. It would be hard to find a blade of submerged aquatic vegetation or an oyster around our ancestral pier nowadays.

Fithian brings me back from my thoughts when he says he hopes that there is "a hell of an oyster season because men crabbing lost five months of this year's salary because of no real income with no crabs this summer. It looks to me like it could be a good season, if somebody tells those boys when they can start . . ." he concludes with another slap at state bureaucrats.

Ronnie Fithian thinks he is fortunate to have a job he likes off the water. He is not optimistic about the future for younger watermen who don't have their bills paid or a nest egg in place. "Oh, there will be some boys that will hold on: Robbie Wilson down on Tilghman and his boys, C.R. and Jason, for example. They know the water, put money aside, and will switch from harvest to harvest if they have to. They can work as a team, the three of 'em, but I'll tell you, with all due respect for C.R. and Jason, I think they'd have a hell of a hard time if not for each other and Robbie. Those young boys without that family support will probably end up workin' the water part-time. I don't see how they can survive otherwise, unless they are lucky and creative."

I leave Ronnie to his zoning and development issues and think about how much he loved fishing. He loved it but he was forced, in his mind, to make a choice between fishing and the way of life he created for himself and his family. He is confident that he made the right decision, although it was not necessarily the one he wanted to make. Perhaps he was right: Only time will tell whether he was a good fortune-teller or not.

LINDA CREWE

For centuries, women have stood by the watermen of the Bay. While only a few of them actually work the water—perhaps fewer than two dozen—their role remains constant. Their primary duties have been to hold the family together during the long hours the men work and to take care of financial planning and management. Traditionally, most men turn their money over to their wives to pay the bills and to prepare for the off-season, but because much of the money is earned in cash, perhaps no one ever knows how much is in a man's pocket—not even his wife. In addition, women assist with net mending, crab-shedding operations, carrying harvests to market, picking up bait, and much more. They provide a strong support system, and this role has changed little over the years.

For example, in Maryland, Brenda Wilson runs the Wilson family's crab-shedding operation and works around the clock at the height of the season when peelers are turning into pricey soft crabs. Terry Spangler crabs with her father, Bill, pulling pots and serving as first mate. Joyce Beck carries her own permits and operates her own boat when fishing with husband Danny. Karen Lassahn works a pound-net operation with her husband John, and Tina Rice handles a soft-crab operation with her husband Billy. Jaimie Dunn and Karen Sue Townsend run their own crab boats out of Virginia.

In my opinion, however, there is not a more active woman on or off the water than the enthusiastic and sometimes controversial Linda Crewe of Newport News, Virginia. I have known Linda for over ten years, and she has never failed to impress me with her drive and determination, whether in a group of watermen or in the government arena. "I know I've stepped on some toes over the years. Men don't like an aggressive woman messing in their work but, hell, a lot of 'em don't understand what the government's doin' to 'em, and I wasn't about to let the culture go down the tubes because of that lame excuse. Me and some other wives took the bull by the horn—maybe more than we should have because I know I pushed too hard from time to time and it pissed people off," she says nonchalantly.

When I met her in 1988, Linda was trying to find a way to unite the watermen of the Working Waterman's Association, for which she served as secretary, with a group she helped start in 1986, Women of the Water, for which she served as president. The women's group was concerned with fund-raising for the men and writing lobbying letters to regulators and legislators. They raised over $30,000 for the Watermen's Museum at Yorktown and additional money to help pay the salary for the president of the WWA while he was off the water lobbying.

Linda Crewe of Newport News, Virginia, sets out to work her oyster and clam aquaculture beds. Crewe has been an outspoken advocate for Virginia watermen for more than a decade.

From 1984 to 1995, Linda and her husband, lifelong waterman Ronnie Crewe, ran their boat, *Linda O'Neil,* out of the York River area. In 1995, when Ronnie's back injuries drove him off the water, Linda ran the clamming boat for a year with a mate and then on her own, until she and Ronnie sold the boat in 1998. That year, Linda started working in aquaculture, growing her own clams and oysters.

Crewe also started lobbying, and by 1999, after years of being frustrated by what she called a "dysfunctional Working Waterman's Association," things finally started improving for her. She spoke in the State House on legislation and regulations and acted as a catalyst in the filing of a lawsuit against the Virginia Marine Resources Commission (VMRC). Funds that had been allocated to replenish the public oyster beds had gone instead to a small group of Eastern Shore oyster aquaculture ventures. The lawsuit was eventually dropped, but the issue created a lot of controversy.

At the time of the lawsuit, Crewe saw what she thought was a good opportunity to regroup. She sent 3,500 letters to watermen around the state to see if there was interest in forming a Virginia Watermen's Association. "I tried for years to get across the success Larry Simns and Betty Duty had in Maryland. Virginia folks just weren't that interested in working together. With testimony in the 'ninety-nine legislature and the lawsuit, I thought the iron was hot. I got 200 responses out of 3,500 and dropped the idea."

But Crewe is a fighter, and while the idea of a statewide unified industry failed, she wasn't about to give up. She studied and researched and single-handedly started the Fisheries Resource Grant Program to make state funds available for people who wanted to test innovative ideas that might improve the industry in any of four major categories: gear, environment, aquaculture, and marketing. Legislation was passed in 1999 to create the program, and $200,000 was allocated for two years. "It's ironic, but while I started the program, I didn't see the legislation pass. I was so damn upset about losing all of our other legislation in 1999 and the fact that nobody seemed to care, that Rob Brumbaugh of the Chesapeake Bay Foundation spoke up for me and the bill passed. He saved it for me and the industry," she says.

Her interest in clams sparked her eight years of participation in the VMRC clam subcommittee and also in the Virginia Commercial Fishing Advisory Board where she helped obtain funds for brood stock to replenish fishery stocks in five river systems. "One of my proudest moments was in 1998 when the Newport News Shipbuilding was applying for a permit to dredge a major clam bed in order to get submarines to their drydock. I didn't think we had a snowball's chance in hell to stop them. I went to the VMRC hearings and ended up stopping them. VMRC Commissioner Pruitt asked what I would suggest doing to resolve this major issue. I told him there were enough intelligent people in the room that we could come up with something. He told me to pick people from his staff, some watermen, and work something out with

Newport News Shipbuilding. If that wasn't something to be proud of, the company agreed to cut down on the acreage they were requesting to just take what they needed. They agreed to hire watermen to remove existing clam beds from that area to move them to one of our brood stock sanctuaries. They also agreed to build a new brood stock area and stock it with seed and chowder clams. I was totally amazed at what we accomplished that day. And a year later we filed our lawsuit against VMRC!

"Now I stay away from all those issues. We won some things and we lost some things. I just grow my clams and oysters and try to mind my own business. When I stepped on toes and crossed the line in the sand with the lawsuit, it caused VMRC and others to turn on me and I was blackballed by people who used to have respect for me. Then I found out that the 3,500 letters I sent out to try to form the new association were misinterpreted by the men because they thought I was asking for money for *me* to represent them, which I wasn't. I was verbally attacked by a waterman for trying to get money for me. He did this in front of about fifty other men and no one stood up for me and what I was really trying to do. The stress of all this, plus broken bones and a hernia from working on the boat, and fibromyalgia, took its toll. They blamed me for their problems when I was only trying to help. I quit all the committees and will leave it up to the museum to preserve their heritage. I guess you can lead a horse to water but you can't make him drink. The time I spent fighting for them for the most part was wasted. Now I'm finishing out my last two-year term on the Marine Products Board and I help out with the Virginia Seafood Council. There is no fight left in me—only pain. That's my story."

Linda Crewe tried to make a difference in Virginia. Perhaps she tried too hard in this subculture where tradition is rock hard and a woman—even though she works the water—is often regarded with skepticism. Perhaps no one will ever know. For now the future may be left to the winds in Virginia because few people are willing to take on the responsibility. Linda Crewe will remain involved in her work and in her family. Unfortunately, she won't be a catalyst for change. There is no fight left in her, and many people may come to regret that.

CHAPTER 7

Back on the Water

As the end of February approaches, I have returned home after fulfilling my NMMA duties at the Miami International Boat Show and the Bay has thawed since air temperatures are reaching sixty-two degrees. Although Congress and the Maryland and Virginia legislatures are back in session with bills in the hopper relating to recreational boating and commercial fishing, I'm anxious to get back on the water. I decide to call Walter Irving Maddox (who goes by Irving) from Charles County to see if he or his son is going out.

After several attempts and messages left on each other's answering machines, Irving and I pick a Thursday forecasted to be warm to join his son, Walter Irving, Jr. (who goes by Walter), and three other men to haul seine mud shad, or gizzard shad as they are more formally known, from the Port Tobacco River. I am scheduled to meet them at Goose Bay Marina on Goose Creek just off the Potomac at 7:00 A.M. sharp.

I'm awakened by a chirpy disc jockey trying to be cute and funny at 4:50 A.M., and I ask Cindy to turn the radio alarm off. She reaches over in a daze and instinctively hits the right button in the dark. We both say good-bye to our chirpy DJ and seek our own peace. Hers is sleep, and mine is the anticipation of going haul seining. I realize now how alive I feel when I know I'm going out on a boat, regardless of the time.

Watermen can't understand why people commute an hour and a half from southern Maryland to Washington, D.C., every day. I'm beginning to wonder, too, even though I know I'll be joining my "bus buddies" on the 6:38 A.M. bus to Washington the next day.

I turn right onto Route 6 in LaPlata and travel a mile or so, and the beautiful homes and farms delight me. As fog lifts slowly from brown-stalked cornfields and a farm pond, light reveals Canada geese nestled by the dozens on the bank. For five miles I pass farms that are picture perfect and I'm envious of the tranquility. Goose Bay Marina and Campground is about four miles off Route 6. There are a couple hundred slips for sailboats and powerboats—all but a few are empty this time of year—and several hundred ad hoc spaces for RVs and campers. It is a nice facility that is entered by a "tarred" road, as Irving described it.

In fact, the smooth asphalt ran at least half a mile through nice farmland with dozens of rolled hay bales wrapped in shrink-wrap, which, until now, I had only seen covering boats in winter storage. Wrapping a large bale of hay in shrink-wrap is like covering it with plastic wrap, only it's white. The bales looked like snowballs five or six feet in diameter lining the side of the road. I thought for an instant that this was a good idea because rolled hay that gets wet rots quickly if it isn't spread to dry out and then rolled again.

As I approach the marina, Irving is sitting in his new white pickup waiting for me. The other boys haven't shown up yet, so he takes this opportunity to tell me about mud shad.

"We call 'em mud shad but the real name is gizzard shad. They stay on the bottom to feed. They aren't good for much except bait fish and when there's a market we catch all we can and they're trucked to Texas or Louisiana for crawfish bait," he explains.

"I started haul seining mud shad about ten years ago after goin' out with a fellow name of Stanley O'Bier from down the Northern Neck [Virginia]. Now he uses bushel baskets with tops on 'em while we use wax coated boxes that are supplied by the buyer. A bushel basket can weigh up to around sixty-five pounds where our boxes weigh about thirty. Pretty good deal if you ask me. Right now we'll get about $3.75 a box or about seven and a half cents a fish, but the market's always fluctuating. Now, it ain't no easy job. Haul seinin' is tough work but it's better than rockfish because there's no limit to it. When I was young I only fished for rock and did a little hand tonging for oysters. That was it. You kind of stuck to what you knew best. Nowadays these young boys have got to know about it all 'cause they have to jump from one fishery to 'nother to make any money. If this market drops out next week then they might go clammin', you know.

"When I started this work up here everybody laughed at me and thought I was crazy. I started makin' good money and now a bunch of 'em are in on it. Why, I know a boy who's gillnettin' down the river for rockfish and he pulled forty-seven boxes of mud shad out of the water on the side 'cause he had a market. Ain't a damn thing wrong with that, is it?" he asks. No, it seems to me, as I have seen so many times working with these men, that Irving's creating this fishery a decade ago is only one example of how creative watermen can be in the hunt.

Irving doesn't haul seine by himself anymore. Just after Christmas 1995, he was nearly beaten to death in his home by two young men, seventeen and eighteen years old. The two are now in prison facing life plus twenty-five years for breaking into Irving's home, trying to rob him, and nearly beating him to death. Watermen are known to keep large sums of cash in their homes before deposits can be made. These boys, one of whom used to work for Irving, took a chance and it fell through. They stole $1,500 from his wallet but were arrested three days later. Irving was sixty-nine years old at the time. He's seventy-five now and "healthier than ever," but he is still in the recovery process because "with a brain injury, it can take years to bounce back," he says.

As we talk, a brown pickup pulls up alongside us. It's Irving's son Walter. We greet him and father mentions to son, with a smile on his face, that he's running late. There is no smile on Walter's face. "Well, we put in twelve hours yesterday and we'll do the same today. I'm just slow rollin' this mornin'," he says. Irving's son is the owner of one of the boats we will use today as well as the haul seine (net), the truck, flatbed trailer, and tractor with a forklift—all necessary to get the job done.

Another pickup drives across the gravel parking lot. Inside are Ernie Woodall, Raymond "Tootsie" Morgan, and Robert "Robbie" Brown, Jr. I know Robert T., Robbie's father. He has one of the quickest mouths on the Bay and can be nonstop if he has Irving present to laugh with him. In the past, I have laughed so hard at Robert T. that tears streamed down my face. Robbie, I will soon learn, is as pleasant and cheerful as his father, and he too will talk a great deal this day! These men come from St. Mary's County near my ancestral home, River Springs Farm.

Ernie asks me what my relation was to Robin Blackistone who was killed a few years back in an automobile accident near the gate to River Springs. I tell him Robin was my first cousin. With that, he shakes his head and says, "He was a good boy. We would shoot pool, play cards, and drink beer." I agree with him and would certainly like to talk about something else, but he's bonding, in his own way, and we move on to discuss Robin's sister Peggy and his mother, my Aunt Sis.

"I'll tell you what," Ernie continues. "I wish I had a dollar for every duck I killed in that hollow by River Springs," he says referring to a shallow cove at the corner of the property where a duck blind sits opposite marsh grass.

The chatter dies down as a bond is formed between us, and it's time to move out onto the water to catch fish. It is decided that I will ride with Irving and Walter. Ernie, Tootsie, and Robbie get in Robbie's boat. These are open twenty-four-foot flat-bottom skiffs. "They fly right on top of the water and can run right good even if they're weighted down with fish," Irving explains as Walter leaves Goose Creek and heads toward the Port Tobacco River. We're on the river five minutes before Robbie leaves the bulkhead, and Walter opens the throttle as we round a point and head upriver.

I'm looking at one beautiful waterfront home after another, all with sloping acreage to the water's edge. Walter points out a large bald eagle perched high in a tree above the water perhaps fifteen feet from his nest.

"He's gettin' ready to go fishin' too," Walter says with a smile. "Got to feed them babies."

Walter explains that he will use a nylon haul seine about nine hundred feet long. One end of the net is anchored to the shore while the other end, with the nine hundred feet of nylon net attached to it, is let out slowly in a large arch and anchored to the shore downriver. Then the net is pulled in, like a shrinking corral, and whatever's in it will eventually wind up in either Walter's or Robbie's boat. The net has now been reduced to a semicircle about 150 feet across and attached to metal and wooden poles driven into the bottom for support and to hold the net about two feet above the water. While haul seining was very popular fifty years ago it's just coming back, according to Irving, and men will haul seine for whatever they can catch: rock, blues, perch, or catfish.

Upriver, we approach the net, and Walter comments on all the seagulls flying over and around the makeshift catch basin. We also start to

Walter Irving Maddox, Senior, *left*, and Junior, head out to collect a boatload of mud shad.

see dead fish floating on the surface as far as a hundred feet from the net. "Damnit!" Walter shouts. "That damn beaver got us again. There's a damn beaver that lives over in that marsh and he chews through the wooden poles to get at the fish."

When we are within ten feet of the net Walter sees a hole in the webbing. He pulls up alongside two poles and asks his father to grab hold of one nearest the bow. He walks up to the front of the boat and the two men reach down to pull the net up. "Yep, he's bitten through the net. Wait 'til Robbie sees this. He'll be some kinda mad. It's always somethin'. Always somethin'," Walter states emphatically, shaking his head. I think to myself, how many times have I heard a waterman say that?

As Robbie, Ernie, and Tootsie approach, Walter raises the net and yells, "The beaver got us again."

"How bad?" Robbie asks.

"Ate through a pole and tore the net here. I don't know how many fish got out but we lost some, that's for sure," Walter explains as the second boat is brought up to the net bow to bow.

Robbie leans over to inspect the hole. "I don't have anything to fix it. Damn, I knew we should have used the rest of the metal poles."

"You try to fix this and we'll go over to the other side and tie up along the shore," Walter says as he moves to the stern and starts the engine.

We move around the other boat, and Irving says, "Robbie's so mad he ain't even talkin'. Now you know he's mad then!" We watch Robbie look for something to use to sew the webbing closed as Ernie and Tootsie hold the damaged section up in the boat.

"Watch in the circle and you'll see a school of shad movin' on the surface when we make a little commotion," Walter says. "They're on the bottom now, in about three-and-a-half feet of water, but you wait 'til we start walkin' around and pullin' some in."

We reach the other side of the semicircle, and Irving thinks he sees another hole made by the beaver. They inspect it and there is a ten-inch opening. They will piece it together later. Dressed in one-piece insulated waders that reach his chest, Walter jumps overboard into about two feet of water. He takes an orange plastic bushel basket out of the boat and places it in the water. "I'll scoop up some of the dead ones and throw them in the boat," he says as he collects dead fish from the shore and alongside the outside of the net.

"You can take them even though they're dead?" I ask.

"These? Oh, yeah." Irving responds. "It don't matter if they're dead. The water's right cold at about forty degrees up here, which will keep them from goin' bad, and they're just used for bait anyway."

Inside the large net, I see several other nets attached to cork buoys, and I ask Irving what their purpose is. "Those are smaller nets, maybe twenty-five feet long which we use to haul seine the fish inside here.

Walter Irving Maddox, Sr., repairs a holding net while standing knee-deep in mud shad in his son Walter's boat on the Potomac River.

Once we do it outside in the big circle then we do it again to get 'em in the boat from here," Walter explains. Wrapped in the smaller nets there are more dead fish that Walter will also throw in the boat.

The men continue to talk about the beaver. "People want to see beavers come back, but all they do is eat as much as they can get in their mouth, dam up the headwaters so fish can't spawn, and tear up everything," Irving states matter-of-factly as I watch Walter take two six-inch nails and place them in holes on the topsides of the boat about five feet apart. "When we're haulin' in the net to bunch up the fish we use these nails to drape the net over so it stays in place and we don't have to hold it from going back overboard. When we gather the seine we call that bunting it," he continues.

Walter wades through three feet of water, picks up two dead fish, and throws them in the boat. "There's fifteen cents," he says with a grin as he moves along the shoreline picking up dead fish and dropping them in the orange basket.

Irving and I use this time to chat, and I value it. He starts talking about eels and agrees with Bob Evans that the state ought to increase the mesh size so smaller eels can escape. "What the hell, you can get $3 a piece for a big eel. Why not let the little ones go?" he says.

"Now if you want to talk about crabs, the Chesapeake Bay Foundation is ruinin' it for us. They should stick to cleanin' up the Bay and not tryin' to regulate us. I do agree with them that nobody should take sponge crabs [pregnant females]. We don't in Maryland but it's still legal in Virginia, North Carolina, Florida, and Louisiana," Irving says.

"Don't take the babies," Ernie says.

"Mick should have been with us the time we had that photographer from the Eastern Shore on board," Robbie says. "We had the boat so loaded down with fish, 237 bushel baskets, he went up to the bow to take pictures of how low we were sittin' in the water. I think he was scared and thinkin' we were gonna sink so he headed for higher ground!"

Irving looks over at Robbie who talks as he finds material to sew the net together. "You're really learnin' to sew pretty good, aren't you, darlin'?" he says across the water with a laugh.

"Got to with a damn beaver eatin' through her," Robbie replies without responding to Irving's jab.

"We'll load up 550 boxes today. That will take care of one trailer, then the boys will come back again and load up another one this evenin'," Irving figures as Walter comes back to the boat. With the net repaired Robbie has brought his boat around the net and stopped behind us. Ernie and Tootsie have entered the water.

Robbie cuts the engine and climbs overboard. "Damn, this water's cold today!" he says.

"Well you're wearing them thin boots," responds Irving. "They won't keep you warm today."

The men in the water hold the top of the net down while Walter pulls our boat over it and into the circle. Then they grab the smaller nets buoyed to cork floats and pull them up to the boat. The men in the water straighten out one side of the net parallel to the boat as Walter lifts the other side in the boat and over the six-inch nails that will hold it together when he bunts. When the fish are bunched together alongside the boat Walter hands Ernie a bushel basket. Ernie dips it in the net pouch and lifts it back to Walter who empties its contents on the bottom of the boat. He then decides to give Ernie another basket so that while he dumps one, Ernie can load the other. Robbie and Tootsie continue to work the net up, while Walter bunts his side when he's not emptying a basket. Once all the fish are out of this net, the process begins with the second small net. The bottom of the boat is covered with mud shad and wherever we walk we step on them.

As the men work, Irving and I watch and talk. He tells me that he has two gillnet stands out in the Potomac River. "I take every fish I can outta there but if you catch the rockfish you've got to tag every one before you leave the net. Now if we're gillnetting for rock up the tributaries you can tag them when they're in the boat but not on shore. And it's a hell of a fine if you get caught doin' it against the regulations."

"That's a lot of regulations to try to keep straight," I respond.

He holds his hands six inches apart and says, "That's how thick the law book is for us. They don't take anything out, just puttin' stuff in. It's a hell of a lot to learn when you're tryin' to move from one fishery to another just to stay afloat. Now they're talkin' about takin' the large rock from us. Why? Old fish don't spawn like young ones. They're just like people or animals. You don't see an old boar hog tryin' to mate when he crosses into old age, or people, or anything. Those big, old fish aren't going to spawn and I can't see why they [DNR] can't get it through their heads. We can take a fish eighteen inches now. Let us get all the fish over twenty to twenty-four inches, that weigh close to twenty pounds, like New York regulations, 'cause they ain't spawnin' and they're eatin' every crab in sight that they can get their mouth on."

As the net is drawn up, it is bunted over the nails that hold it in place. Three men stay in the water with the net and basket while Walter stays in the boat still dumping fish. Robbie, in his thirties, is talking a mile a minute and making the men smile and Irving laugh. He's talking about the cold water and his new hip boots. "I'd rather my feet were cold in these boots than my whole body goin' overboard. That flat bottom boat will throw you in a minute if she hits a wave. Now Tootsie doesn't have people fall off his boat. He throws them off," Robbie says as Irving starts laughing.

Tootsie talks about the time his pregnant wife and son fell overboard. "He did a quick learn for a small boy that time," Tootsie says.

They all tell stories of falling overboard until Walter sees a school of shad on top of the water at the far end of the net. The catch nets are empty so he tells the three men that we're going over to the school and they should spread out their net and start walking toward the fish. His intention is to drive the school into the net the men are holding. "We'll make a disturbance by revving the motor and try to drive them like cattle over there so they can corral them," he says.

We are barely moving when he throws the gearshift to neutral, revs the engine, and puts it back in gear steering the boat along the edge of the school to maneuver them to the waiting net. The men keep moving and now they are about fifteen feet from us with the school between them and the net along the perimeter of the circle. The fish slowly get trapped and Earl and Robbie draw it closed as Tootsie brings the other

Ernie Woodall, *left*, and Robbie Brown, Jr., stand in the river to collect the net loaded with shad.

small net over to catch more. These watermen are looking more like cowboys every minute!

"Damn if you boys ain't got a load in there," Irving says with a laugh as Robbie and Earl pull their net back across the water to Robbie's boat. Tootsie pulls his net together and ties it shut so it's like a big bag and follows the other men.

By the time they are at Robbie's boat, which is outside the perimeter, Walter has our boat parallel to it. He bunts one side of the net to hold the large catch in place while the men pull it in as far as they can. Once the net is as bunched as they can get it, Earl steps inside of it with a bushel basket and starts handing full baskets to Walter to put a few more fish in his boat before they start filling Robbie's.

With the boat full, we are forced to walk in an unpleasant combination—bloody water three inches deep and fish two feet deep. Walter asks me to join his father in the bow of the boat, so I start pushing my boots through fish in order to find the bottom. It's impossible, so I settle for walking on them. Irving warns, "Better hold on to the topside there. You don't want to fall in that mess." It's not an easy task to walk on hundreds of wet fish for fifteen feet without holding on to anything. I almost trip over the catch trying to reach the side of the boat, and Walter starts laughing. When I'm secure sitting on the topside in the bow I ask Walter how many fish they will catch from this haul seine.

"All told we'll get about 4,500 boxes out of this net from when we brought her in last week," he explains.

"We've put eighty bushel baskets in here and that equals about a hundred boxes," Earl adds. With all the commotion he was still responsible for keeping a relatively accurate count on how many fish they will take from the net today. Walter wants to fill two trailer quotas but he certainly doesn't want to waste any fish by having them on shore with nowhere to go. They put eighty baskets in Walter's boat and then switch to Robbie's.

"Besides, with a bit of a sea this mornin' we don't want to overload. With eighty in each boat plus what we have on shore, that will fill up the trailer. We'll come back this evenin' and put eighty in each boat again and that will take care of the second trailer," Walter says.

As fish are emptied and the net is bunted between the boats they are drawn closer and closer together. "This is a load of fish but that beaver is killing us. This is the second time. Tootsie says tonight's the night," Robbie says to anyone listening.

"That's it. Tonight's the night for him," Tootsie confirms as Walter laughs.

"I'll give you twenty dollars if you stay up here tonight and take care of him," Robbie says to Tootsie.

"I'm stayin'," Tootsie says.

"I'm stayin' too then," Robbie responds.

"Tootsie" Morgan, *left,* and Ernie Woodall gather the net so Walter Maddox, Jr., *in the boat,* can tie it off to bring the shad closer to their scoop nets.

"If you get him you can barbecue him good," Irving says.

"I don't want to eat the damn thing," Robbie says to Irving, "I just want to kill it!"

With the nets emptied, Walter wants to repeat the whole process. The three men start walking across the large corral, spreading their nets as they go. "Let's go round 'em up again," Walter says as he slowly moves the fish toward the catch net. "This is the first time we've tried this roundup approach and it works pretty good."

"Why don't we see other species in the net if you took everything out there when you laid out the haul seine and brought it in to shore?" I ask.

"Oh, they are there but they're on the bottom and will come up with the last batch of shad. Sometimes there aren't other fish. We've had a couple cats already today but we turn them loose."

Once again, with full nets, the boats are brought together; nets are hung over the nails and are bunted by the boys in the boats.

"Ain't nothin easy about haul seinin'," Walter says, "but as long as we have market and we can haul fish it's as good as doin' anything else." I guess men have said just about the same thing for over two thousand years.

With two boats filled with the required amount of fish, we head back to Goose Creek and the marina where we'll unload the catch. It is a beautiful day on the river as the sun sneaks through cloud breaks to reflect off the water in an unseasonably warm sixty-nine-degree temperature. Gulls remain everywhere in the sky and on the water, still feasting on the dead shad floating with the tide and current. Downriver I look at the skeletal tree with the large nest built in its highest chamber. The eagle has left for hunting and I hope he goes upriver and takes shad laid out buffet style, compliments of a beaver.

Back at the marina both boats are tied up parallel to the bulkhead. Tootsie and Walter go get a flatbed trailer as Robbie and Earl put in the

A boatload of mud shad that will eventually end up in Louisana as crawfish bait.

boat about twenty wax-coated boxes, perhaps two feet long by eighteen inches wide and five inches deep, and start shoveling fish in them with large coal shovels. As the boxes are filled Robbie lifts them to the top of the bulkhead.

When the flatbed trailer is backed up parallel to the boats, Tootsie begins carrying the boxes to a wooden pallet he has placed on the flatbed. The pallet will hold five boxes. Once five boxes are laid out in a square, a wooden piece of lattice is placed on top of the open boxes of fish and five more boxes go on top of them. He will stack them five high, twenty-five boxes to a pallet.

As boxes and pallets fill it still seems to me that the shoveling makes little dent in the amount of fish in the boats. "I bet you're glad to see the deck, hey Robbie?" I say with a smile.

"Damn right," is all he says as he and Ernie keep shoveling and then placing boxes on the bulkhead. At this point, sense of pride and appreciation lead me to leave my oilskins on and help Tootsie with stacking the boxes on the pallet. I almost drop my first one, not realizing that the waxed cardboard, when a little wet with fish scales and grime, can be very slippery. Raising my right knee quickly under the box for support, I release my grip and quickly move my hands beneath the box making it easier to carry and load. As we're loading the last boxes on the last pallet, Tootsie, in heaving his box on the top layer of a pallet, lets it slip from his grasp. Fish fall all over him from his face to his feet, landing in the sand gravel drive.

"Damn it!" he yells, brushing fish scales from his cheeks and T-shirt while the rest of us nearly fall over laughing.

"The fish that got away," Robbie says laughing, as he and I kneel down to start picking up the fish and replacing them in the box. "This box will be heavier now that rocks are stuck to it," Robbie tells Tootsie with a final jab as I pick the box up and heave it on the top of the stack.

My right elbow and forearm muscles are killing me from tendinitis. I was about to say something while lifting the boxes to the pallet but a quick flashback and self-preservation prevailed. A few weeks before, in the company of unnamed watermen, I mentioned the pain without

thinking and was soon the victim of much teasing. The advice and offers
to help I received were accompanied by a lot of laughter and little sym-
pathy. I had made a crucial mistake. I opened myself up to jabs and
spontaneous joking, a trademark of these men. I would not do it again
here. Especially with the way I knew Robbie, Irving, and Tootsie would
probably react to such a simple admission. Let Tootsie, with fish scales
on his eyebrows, receive their wrath today—at least until he found his
other clean, dry shirt.

With the first flatbed loaded with six pallets of fish, Walter gets a
smaller flatbed trailer and parks it near pallets on the ground that we
have loaded five high. These four pallets contain the last of the fish from
the two boats, a total of two hundred and fifty boxes of fish. At about
thirty-five pounds per box at a little over seven cents a pound, I'd say this
is not a bad catch for the first of two nearly identical trips that will take
place today.

Tootsie Morgan loads boxes of mud shad on a trailer. They will go into a
refrigerated tractor-trailer to be carried to market.

As we load the last of the fish, Walter gets his forklift to pick up the pallets and place them on the smaller trailer. Once that is complete, he tells Robbie and Earl to drive the truck with the large flatbed and Tootsie to drive his truck with the smaller flatbed up to the trailer parked at the end of the property. Irving and I drive our own trucks up behind the fish caravan and forklift for a quarter of a mile to the trailer.

Once at the trailer, Robbie climbs up on the front end where the refrigeration unit is located and turns it on. At the same time, Tootsie opens and climbs into the back of the trailer and moves a small hydraulic jack near where Walter will place the first pallet once he's gotten it off the flatbed. As each pallet is placed in the trailer, Tootsie slides the jack under it and puts it in position next to or in front of the last one. Soon the trailer will be full of pallets containing 550 boxes of mud shad, locked up and ready for the buyer, Bevens Seafood of Virginia, to take the eighteen-wheeler down the road from Goose Bay to the Northern Neck.

Walter is finished unloading and walks up to his father and me. "Well, we'll get some lunch and go do it again."

"Hey, Walter," Robbie yells from the front of the trailer. "We got a problem. A fan belt broke and the refrigeration unit won't work. She cuts off when she reaches 160 degrees for some reason."

"So it won't burn up. Let me see it," Walter replies as we all look at a fan belt that probably stretches six feet. "This is going to cost probably forty dollars to buy a new one and then we've got to fix it."

"I wouldn't know where to get one and I'm not spending my money on it." Robbie responds.

"Well, go call Bevans and tell them we'll have two trailers loaded tonight but we've got a busted refrigeration unit," Walter instructs as I prepare to leave them.

Bevans Oyster Company of Kinsale, Virginia, buys mud shad, freezes it, and ships it to Texas or Louisiana. With few small packinghouses left on the Bay, Ronnie and Shirley Bevans are a blessing to many watermen and to the local economy. The company started with six employees and one acre in 1966 and has grown to over two hundred employees in several locations. The company's ability to plant and harvest its own oysters has been a crucial component to its overall success. The Bevans Brand oyster has become well known across the United States.

I'm confident as I head to my truck, after shaking hands and thanking each man, that they will be able to finish the day without me. As I'm driving by, Robbie stops me and says I'm welcome to go clamming with him in a week or so, since they probably won't have market for shad. I thank him again and leave them staring at a torn fan belt and wondering how to keep the fish at forty degrees if they don't get a belt soon.

MORE WINTER FISHING

By mid-March the Maryland and Virginia legislatures are in full swing and watermen in the Bay region add another layer of anxiety to their occupation. When the legislature is out they can concentrate on harvesting seafood or mending equipment and can watch from the sidelines as Larry Simns and others in the Maryland Watermen's Association monitor regulatory and legislative discussions. When legislatures are in session, proposed legislation is often dropped in the hopper and a hearing scheduled. On a moment's notice, they must be ready to leave the water to head to Annapolis or Richmond to be heard collectively on a wide range of issues. While Larry Simns takes care of most legislation through lobbying techniques that often match the highest paid lobbyists in the state, there are times when watermen are asked to help.

I remember when the rockfish (striped bass) moratorium was being lifted, hearings were scheduled in Annapolis to debate whether striped bass should become a game fish. The watermen came in droves to testify and so did the sports fishermen. So many came that the hearing was moved from the Environmental Committee room to an auditorium in the State House. It was an interesting sight to see the room packed on one side with men in flannel shirts and white rubber boots and on the other side with men in khaki pants and docksiders. The room was divided down the middle by state employees from numerous departments and curious onlookers eager to see the watermen at work. In truth, many came to see Larry Simns as he went up against game fish proponent and lobbyist Bruce Bereano, the highest paid state lobbyist in Maryland—perhaps on the East Coast—with annual fees in the millions of dollars. Simns won the round that day.

With such memories lingering in my mind but knowing the legislature was not expected to see heavy combat, I called Danny Beck to see if he was fishing. Beck is from the tough, blue-collar industrial area of Essex and Middle River just north of Baltimore's waterfront industrial plants. He is respected around the Bay as a fine waterman with a reputation for speaking his mind, regardless of who his victim might be. Like Bob Evans, Beck, as president of the Baltimore County Watermen's Association, is often called on to testify on specific fishery issues, particularly if Larry Simns is traveling on other MWA business. While perhaps as well versed on the issues and fishery concerns as Simns, Beck is, by his own admission, not the politician or diplomat that Simns is. Nevertheless, he is quite adept at getting the job done, either on the water, at a regulatory agency, or in the State House.

Danny sounds glad to hear from me when I reach him on March 15. We had talked several times over the last several months and he had promised to take me out. He is going to pull fyke nets at six o'clock in the morning and he invites me to join him. After giving me directions to his house, easily two hours from mine, he closes the conversation by saying simply, "Follow those directions I gave you and you can't get lost. See you at six." I immediately know two things to be true: First, he is in a good mood, which is the only time I want to be around him, and second, I'd better not be late.

In the dark, I find his street sign and follow a dirt and gravel road for about a quarter of a mile until I see the sign that says "Beck's Crabs." I turn right and pull into his driveway, which winds its way around several thousand crab pots, piles of pound net poles, several trucks, a forklift, refrigerated buildings, work and storage sheds, an ice machine, countless gas cans, a couple of dumpsters, and five beautiful geese hand-raised by Danny's wife, Joyce. I park down by the pier next to the only other car in the lot. I assume this is a member of the crew resting in his seat. Everything is quiet as dawn breaks and a calm, clear morning waits impatiently for us to move. I sit in my Jeep waiting for noise or movement either from the car next to me or from the house where one light shines in the kitchen window.

Finally, I get out. I need to relieve myself of the 7-Eleven coffee I picked up on the way to Essex. My movement draws some attention as a small, white-haired fellow steps from the car beside me.

"Mornin'," he says as he offers his hand. "You workin' today?"

"No, I'm writing a book on the watermen of the Bay and tagging along with you guys today."

"A book. No kidding. Well, I'm Dan Fitzgerald. You goin' out with Danny will give you plenty for that book. He's the best man I ever worked for. He's smart. Knows the river. And has his opinions of everything," he says with a slight smile.

"I know. That's why I want to go out with him," I say with a grin in return.

Within minutes a white pickup drives up and a lanky waterman with long blond hair steps out. He introduces himself as Joe Rohlfing, or "Ick." Dan tells him I am writing a book about watermen and today it was about their fyke netting. I immediately have another new best friend.

As I survey the surrounding gear ranging from seine nets to crab pots, Joe begins filling the gas tank on a twenty-foot open boat. There are two other baybuilt boats in the forty-foot range tied up to a pier and bulkhead parallel to the open boat. The big boats are Danny and Joyce's that they use for crabbing. They both have licenses to crab so that they can double their market in a good year. I am immediately struck by the purple hulls and recall that purple is Danny's signature color: purple shirts, purple hats, and purple hulls on his boats.

After Joe fills the gas tank, he moves to fill the tank on a forklift parked ten feet from the boat. "So you guys are pulling fyke nets today?" I ask.

"We'll pull about six of 'em and then formulate a game plan," he responds as Danny strolls from the house toward the dock. He is a big man dressed in three layers of sweatshirts and fleece, sweatpants, and a black stocking cap with the words "Caterpillar Diesel" in yellow on the front. A gray speckled beard covers most of his face. If it were white he might pass for a Kmart Santa. We greet each other with a handshake as he steps up into the seat of the forklift. The morning has begun.

Dan and Joe load a dozen plastic baskets on the boat as Danny tries to start the forklift. It won't start, and he tells Joe to remove the engine cover as he steps from the cab. He goes to the fish cleaning station on the pier and returns with an empty plastic Coke bottle. After pouring about four ounces of gasoline in the bottle he hands it to Joe and instructs him to prime the carburetor just before he turns the engine over. On the third try it starts and they decide a new fuel pump is in order.

Boat hooks, baskets, and four-foot-long stakes are put on the boat. The baskets will hold yellow perch, catfish, bigger white perch, and maybe a few other fish that we don't want to sell but are good to eat. They will put fish in them that we come across on the trip to pick up the nets. The fish will be separated as they are pulled from the nets or hedgerows. "Kind of like cullin' oysters, only in this case it's fish," Danny explains.

We move out of the creek toward the Gunpowder River. Hart-Miller Island rests off to starboard. The morning sun sends a soft orange glow from the east but it can't counteract the biting breeze hitting us in the face as Danny opens the throttle to reach his nets several miles upriver.

As we watch six white-tailed deer bound up a slope from the river that borders Aberdeen Proving Ground, a huge federal installation, the topic of conversation is Joseph Palczynski, an armed survivalist who murdered four innocent people after an argument with his girlfriend. Palczynski, the police believe, is somewhere in the massive acreage of the proving ground or the adjacent Edgewood Arsenal. Police, on foot, in cars, and in helicopters, are searching the area with a fine-tooth comb. Unfortunately, the killer is equipped with survival gear, weapons, and perhaps more importantly, a battery-powered television so that he can observe the information the media is passing on to an anxious audience.

Dan rests on the floor of the boat barricaded from the wind by the bow, while I sit on a beam next to the steering column within earshot of Danny. "Mick, you know where that guy is that killed all them people? We're going right where they think he's hid out," Danny explains as a police helicopter, which broke the silence minutes before, appears overhead about a mile away.

"Yeah, I told Joyce if he comes around the house give him a little dose of the elephant gun under the bed!" he says, as Joe, who is sitting on the large aluminum holding box, moves closer to Danny. "He's makin' an ass out of the government. They're using heat sensors from the helicopters and we just saw six deer run through there. What rocket scientist bureaucrat is going to tell between him and a deer? I'll tell you what. If he had a pot seed on him they could bring their dogs to sniff him out," he says with a grin.

We are going up the Gunpowder River that Danny has fished all his life. He has market for white perch because the yellow perch are "done runnin'." "You can't keep fish in February. The state season starts on March 1. If we even *see* a yellow in the nets we have to let 'em out or it's a crime. The problem for us is that they spawned early this year so they were about done when the season started. Two weeks in March is it. Once they spawn they disappear. The yellows we are pullin' out of the nets today have been in there for a week," he says.

On this first run up the river we are not emptying nets. Danny wants to check them to see which ones need to be emptied and wants to reset others. Fyke nets are slightly different from the hoop nets used for catfishing. To set them, you anchor one end near the shore and drive in two stakes on either side of a one-hundred-foot-long net to hold it up in the water. The net is called a hedge because it looks like a three-foot-high hedge around someone's lawn. The hedge runs into the fyke net, which is a twenty-foot-long "barrel" wrapped around a series of hoops and has two wings that extend to the sides. A three-foot-long stake then anchors the net. Dan explains that "Danny designs his own fyke style, like he does with most things around here. He wants certain things to happen so he puts a twist here and a tuck there. He's tryin' to outsmart the fish. He's good at it, too."

The theory is that the fish bump into the hedge and follow it all the way up to the fyke net. In reality, some fish are often gillnetted in the hedge as they try to get through rather than follow it up to the fyke.

I ask what the market is for perch. Danny says right now it is between $1.50 and $2.00 a pound for yellow and between $.40 and $.60 for white. They could end up in Canada or California. He will ship the perch to a wholesaler in the Jessup, Maryland, seafood warehouses.

As we reach the first net, marked by a small cork float bobbing on the surface, Joe reaches over with a boat hook and pulls up the anchor stake. Danny puts the engine in reverse, pulling the line taut. The stake is driven back in the muck with a makeshift pile driver that he uses by hand. As the net now reaches close to the surface I see one white perch in the hedge net. Joe removes it. Danny tells Joe to wash the mud from the gunnel. He doesn't want it to get on the net and foul it up.

Moving to the next net Danny says, "It's goin' to blow like a mother tomorrow 'cause it's the seventeenth of March. It always blows like a bastard on St. Patrick's Day."

Dan and Joe are pulling the nets Danny wants to take back to shore to power wash. As they pull up the hedge Danny notices a piece of seagrass and sees that the net is twisted as it is about to go in the boat. He wants the grass removed and the net straight.

Fish in the net are pulled and either tossed overboard to the scavenger gulls now around our boat or put in the holding tank. The fyke net finally reaches the boat and it is folded easily as the hoops drop to the floor. Joe then pulls a slipknot, which secured the anchor line to the net. Danny takes the line that is now only secured to the anchor stake. He wraps it around a cleat fastened to the transom and puts the boat in reverse until the stake is pulled loose. He puts the boat into neutral and pulls the wooden stake from the water, dunks it several times by its anchor line to wash the mud from it, pulls it out, wraps the line around it, and tosses it in the corner by the gas tank. This process will continue until he pulls all the nets he wants to take back in: pull hedge, fold, pull fyke net, fold, pull stakes, wrap lines, repeat. They do this over and over, day after day, throughout the entire season.

As they work, Danny talks openly about a variety of subjects. "Gettin' rid of watermen ain't no science. It's rich people who want me out. Money baggers tryin' to outlaw us because I have a strong relationship with the legislature and DNR. The administration of the DNR is environmental, academic people who don't want to understand what we do. They don't want to pay attention to good science and reasoning. They want to listen to the developers and waterfront residents."

We come up on the only other boat in the river where two men are culling fish. Danny and the men exchange greetings. He tells them he's pulling his pots (nets) because the temperature is rising and fish are going over the top of them, which they tend to do to get to warmer water.

Fyke nets are cleaned with high-pressure hoses to remove algae and mud that accumulate on the nylon netting. Continual maintenance of boats and gear is a necessity when working in salt water or in a harsh environment. Courtesy Maryland Watermen's Association.

The conversation shifts briefly to the killer running loose on shore. "There are so many cops they're chasin' deer everywhere," he says as we pull off and head for another location where he has nets staked out.

With the wind hitting him square in the face he pulls up his sweat-shirt hood and snaps his foul-weather jacket near his neck. My back is to the wind and with my hood up I'm relatively comfortable and in a position to listen. Without reservation I decide to let him talk and forget about my questions. He begins to talk about the rockfish moratorium of 1985. As a member of the Rockfish Compensation Program, designed to give financial help to fishermen, he has his own opinions about one of the biggest issues to hit the commercial fishing community in decades. "See, DNR said the 1983 class of fish, based on their survey, essentially resulted in no hatch. Well, when I fished the river for the state survey people there were so many little rock from the '83 class it was unbeliev-able. DNR said they didn't exist! I called Channel 11 TV news and took this reporter out with me to show him the damn fish and let him get them on film. He got it on film but it was never seen.

"I was mad at Larry Simns about all that bull too. I never trusted him for a long time. I didn't care what anybody else said about him. When I was president of the Baltimore County MWA I tried undermining him every way from Sunday to test him out. I told him I did it, too. I told him, 'Simns, I've stabbed you in the back so many times it's amazing you ain't bled to death yet.' He looked at me and said, 'I'm still standin', ain't I?' Hell, I was trustin' him with my livin'. Since then, I ain't mistrusted him a hair. He's a hell of a man, Larry is. I feel sorry for him sometimes. He's givin' his heart and soul to the watermen of this Bay and, just be-cause he's involved in a decision that doesn't sit well with some boys, the SOBs go and put sand in his gas tank or cover his boat with manure. Some days he can't win for losin' but he sticks with it—for us."

Danny says that in the 1950s and '60s his father was president of the Watermen's Association "and it was the same then as it is today."

"What will happen when Larry retires?" I ask.

"Well, I'll tell you. Men from all over the state have talked to me about runnin'. I'm talkin' about fishermen, crabbers, oyster boys, all of 'em. But I watch Larry and I talk to him. It will take a different kind of man than me. He's travelin' all over the country on fishery issues tryin' to protect us all. And he's had to kiss every political ass between Canada and the Caribbean. He's had to because that's how you work the politics. I won't kiss anyone's ass! We're talkin' about a few boys that'd probably be good. I think Kenny Keen is one for sure."

I comment that I like and respect Kenny.

"Well, I didn't know him too good but I knew he helped Larry and the association a lot. Then I was talkin' to him at the convention in Ocean City. We were talkin' about my lawsuit over fish license transfers and quotas with the feds and maybe goin' to jail and all and he looks me

Danny Beck is dressed to protect himself from the wind and the rain on a good day for fishing.

straight in the eye and says he will help me any way he can. I knew then and there that he was a good man."

"Workin' out here means commitment. I'm ornery, I know. But I'm committed to this industry and the men and women. My will says if Joyce and I die at the same time, 25 percent of all my estate goes to my mother, 25 percent goes to the Zion Lutheran Church where I was raised, 25 percent goes to our kids, and 25 percent goes to the Maryland Watermen's Association for a health and welfare fund. Now I don't talk about that, you understand, but by God, that's commitment. And we need more of it rather than whining or crying because you don't get what you want from Larry all the time."

Danny Beck leaves little to the imagination. As he talks between nets, Joe moves up to listen. Danny is not ashamed of anything he says because it is how he feels, and for him it is the truth.

I am reminded of my conversations with Irving Maddox when Danny says, "You can have laws and regulations—that's not the problem. The problem is in the interpretation and who interprets it. You need a damn law degree to be a waterman if you don't want to be busted for somethin'. Hell, watermen always have a light out, life jackets where we can't get at 'em, somethin' they can get us for. The state boys are pretty good and leave us alone to make a livin' in that respect. Just don't play too many games with your catch."

Danny wants to revisit the rockfish moratorium. "Back in September of 1984 the state offered up a 55 percent reduction in catch before the moratorium went in. They called a meeting here at the Bowley's Quarters Improvement Association to talk about the proposal. What no one knew was I had a copy of an in-house DNR memo, dated September 11, days before the meeting, signed by the DNR Secretary Torrey Brown, saying there would be an outright moratorium, which meant this meeting was a damn sham.

"Then when he and Verna [Verna Harrison, then head of Tidewater Administration, the division of the Maryland Department of Natural Resources that regulates fishing] sat up in front of 295 people who thought they were there to express themselves and negotiate about a 55

Joe "Ick" Rohling unties a
fyke net to spread it out for
fishing off Aberdeen Proving
Ground in Baltimore.

percent reduction in catch, I listened with everyone else. Finally, I had
enough and I raised my hand. In front of the whole place I read the
DNR memo and you could have heard a pin drop. Then I said, 'Dr.
Brown, see this two-inch lump on my back?' I asked as I pulled my shirt
up. 'If I have cancer I'm goin' to take you to hell with me because you're a
liar and I don't want someone regulating me for special interest groups.'
The place went wild with cheering. I said what I had to say. Years later I
apologized to Torrey. And I told him I didn't have cancer. He told me
he was damn glad!" Danny grins.

"I didn't grow up with this attitude. Some guy in Annapolis gave it to me. I speak my mind. Most guys out here won't do that and half of 'em, I swear, could go postal at any minute because we're all facing the same stuff. I just don't hold it in. I'll tell you Pete Jensen should be secretary of DNR in my book. He knows and cares about the resources. He's 100 percent fair to all the user groups and he has everybody's respect. You can take that bunch they have in there now and run 'em up the flag pole.

"I used to think the biggest problem we had was pollution. Well, the Bay's improvin' and they're still after us. But the biggest problem we have is groups like the CCA [Coastal Conservation Association], a bunch of volunteers who don't want us to catch neither fish. And they push DNR for more regulation and legislation.

"Listen, we have cold water algae on our nets which comes from fecal coliform. It fouls the nets up and the fish won't go in 'em. DNR said it was human, from septic overflow. I believe black ducks are the main problem. A few years ago, they wanted to close the Wye River because of human waste and it was goose shit! We have millions of dollars in sewage treatment plants all over the goddamn Bay and the shit's back with vengeance. It's the proliferation of ducks, swans, and geese. And I've got to pressure wash all the nets all the time because of them."

We have pulled three sets of nets as he continues his monologue.

"Now there was all this bull goin' round about pfiesteria in the last year or so. Fish with lesions, men with lesions, close the rivers, don't eat the fish. Pfiesteria has been around for six hundred million years. Menhaden and croaker act it up. The massive fish kills on the Bay when you and I were kids were attributed to lack of oxygen and algae blooms, which is true sometimes, but we'd catch young alewives with their tails eaten off and I think that was pfiesteria. Yeah, you can blame Frank Perdue and all his chicken farmers for runoff of waste and fertilizer but I don't necessarily agree that they cause pfiesteria—unless he's one old SOB!"

I interrupt him to ask his opinion about the foreign crab meat issue and Bob Evans's concern over Phillips Seafood buying their meat overseas. Legislation to require labeling of cans, identifying domestic or foreign meat, died in the environmental committee but Phillips, DNR, and watermen all agreed to forming a task force to address the issue.

"Listen, Steve Phillips has more money than the Catholic Church. He invested his money overseas to create a resource to meet his needs that couldn't be met here. Our packers have a problem with that and so do some crabbers. In 1997 I only got $8 a bushel for crabs, and the Bay packers need to realize modern technology delivers quality meat and it's easy to prepare. Very few shells.

"I've known Steve Phillips for a long time. He's a Hooper Island boy whose family built the empire. It's better than some foreigner with a lot of oil money coming over here to change the industry. Hell no, I'm not mad at Steve Phillips. I'm proud of him and what his family has earned. They've expanded the market and put confidence in seafood!

"When all this came to a boil last fall, a bunch of us went to the Phillips plant for a tour on December 17, 1999. They were shovin' out 100,000 crabcakes a day, and 20–26,000 pounds of crabmeat a day. They package all their meals, frozen food, and restaurant food right there in Baltimore. He's got three hundred people working in a plant that you could eat off the floor of and he could take every crab out of this river and the Bay and still not have enough. That's how I feel about Steve Phillips."

About fifty yellow and white perch come out of the third net pulled. Joe secures the anchor stake line to a cleat on the bow. Danny revs the engine and pulls the stake out. They repeat the same process pulling the rear stake from the water. Joe and Dan then begin pulling in the fyke barrel net followed by the hedge net.

As the net is secured on top of the others in a neat pile, we head for Danny's dock on Brown's Creek. He has lived and worked on the creek all of his fifty-two years. With the throttle wide open we all bundle up. Joe moves in so he can hear more of Danny's thoughts to take advantage of the opportunity to learn another side of his captain and get a new perspective on the industry he is depending on for his livelihood. Although the opinion may be prejudiced, it is fascinating to both of us, each for our own reasons.

"Take horseshoe crabs," Danny starts again. "When the water's cold they don't crawl but when the air temperature heats up and the birds fly up from South America or someplace to eat the horseshoe eggs DNR says there aren't any eggs. They see the damn birds on the beach and still say we've overharvested to the point that there are no eggs to speak of. Like Bob [Evans] told you, we can't take horseshoe crabs now, even for our own bait. We've got to go to Delaware and buy them! They can piss me off with the way they think!"

I stare at the white fishtail of water rising four feet from the surface in a whirlpool of action trailing the boat and spraying a fine mist to either side. It's mesmerizing. I stare and think about all that Danny Beck has said and realize when we slow down to pull into Brown's Creek that the day is not half over yet. There will be more to learn.

As we approach the dock a blue heron standing three feet tall is on a piling at a neighbor's pier. There is no activity on the waterfront; the summer cottages are vacant. Lawns are mixed with hues of green and winter brown and dotted with onion sprouts. Several mallards swim lazily between isolated piers, and the heron watches.

"That's Danny's heron," Dan says matter-of-factly.

"What do you mean, 'Danny's heron'?" I ask with some astonishment at the statement.

"Watch this," he says as we turn to Danny who has climbed out of the boat and is standing halfway down his pier with a seven-inch white perch in his hand waving it like a member of a race driver's pit crew trying to get his attention.

"Come on, old man," Danny shouts waving the fish in the air. "Come on, old man," he repeats five or six times as the heron watches.

"That bird will come over and take the fish right from his hand. He's here every day when we come in and every day Danny calls him and he comes to get the fish," Dan says as Joe nods his head in confirmation.

There is work to be done and we must get back on the water. Danny tosses the fish onto the pier and strolls back to the boat.

"He'll get it while we're gone. This afternoon, when we come in for the day, he'll take one from Danny's hand when he calls him. The heron has been doing this since 1979. A week after Danny's father died the bird showed up and he's been here every day since. They say when a fisherman dies they come back as a blue heron, but I don't know for sure if that's true," Dan informs me as I think that not fifteen minutes ago Danny was swearing and angry about countless issues. Now he is lost in the peace and compassion he has for an old heron that he calls "old man." Perhaps the saying is true and, in this case, only Danny knows for sure.

Danny goes to get the forklift to pick up a pallet with tall stacks of plastic tubs on it. Each tub holds a hundred pounds of fish. Joe and Dan place them near the bow.

Next, an aluminum culling board twelve feet long is carried from the shore and set parallel to the port side. Two large plastic trash cans are put on the boat. They will support the culling board when it is in use. Finally, Danny returns with the forklift holding an aluminum holding box, perhaps four feet deep by four feet wide and seven feet long. He gently sets it on the gunnel as Joe and Dan slide it into the boat and push it back against a bulkhead that separates it from the steering column. It has a wooden top that is tied shut until we get out to the nets. There is little room to sit. Dan moves to the bow of the boat between the plastic tubs. Joe and I sit on the holding box facing the stern.

Leaving the creek I again stare at the three-foot-tall rooster tail, trailing thirty feet behind the 175 horsepower Mercury outboard, as a ray of sunlight fights its way through overcast skies on the long run back out to the nets. For all the talking and listening we did while checking and pulling nets earlier, the trip back out finds each man lost in his own solitude. Iridescent fish scales, like quartz shavings reflecting that lone ray of sunshine, are everywhere from stern to bow, on the engine cover, gas tank, baskets, and our foul weather pants. It's nice out here.

Approaching the first buoy Danny pulls up on the throttle and reaches past me for an end of the long culling board. Dan reaches for the other end and pulls it on top of the holding box after Joe has raised the top and cleated it off in front of the steering column. They move in unison, in tight quarters, with me conspicuously right in the middle of where they are trying to do most of their work.

"Just stay right there, Mick. You're okay. We'll just move around you and then we're set," Danny says, making me feel more comfortable.

When Joe pulls up the float buoy, Danny takes the line and cleats it off next to the steering column. He revs the engine in reverse for a split second to pull the anchor stake from its hold on the bottom. We move the boat down the hedge until we reach the first hoop of the fyke net. Danny has left the wheel and relocated to the bow with Joe next to him. Dan has moved from the bow and now is between the holding box and the starboard side of the boat wedged in about a two-foot space just in front of the steering column. Danny reaches down from his spot and grabs hold of one of the hoops. As he lifts his open end to the gunnel, Joe is pulling up the rest of the net with help from Dan. This moves fish to the front of the barrel. Once there, Joe takes hold of the front hoop; that frees Danny to throw a floating net basket into the water. Shoving a foot of netting from the front of the barrel into the floating net basket ensures that the fish he and Joe are pouring from the fyke do not get away.

Once the fyke net is empty, Dan grabs a dip net, or crab net, while Danny and Joe slide the culling board on top of the two large trash cans. When the board is in place, Dan takes the dip net and reaches over the side to scoop fish from the float net. He pours them on the culling board as Joe and Danny move quickly to sort the fish. The process is repeated until Dan takes the last of the fish from the float net.

Small catfish are shoved off the open end of the culling board back into the river, along with a large carp, several nice sized rockfish, black sea bass, and herring. White perch must be at least eight inches long to keep; yellow must be eight and a half inches. There are fine lines etched in the culling board for each. When the men aren't sure if a fish is long enough they hold the head against the side of the board and push the tail down with their fingers. If it reaches the correct line it goes in the holding box if it's white and in a tub if it's yellow. Smaller perch go over the

Danny Beck culls his catch before putting the fish in baskets to be iced down for market. Courtesy Maryland Watermen's Association.

side with the other discarded species. Very few are dead when they go overboard. Those that are will be quickly snatched from the blue-gray surface by dozens of hungry gulls circling overhead. A heron swoops down among the gulls and leaves with a small white perch.

Joe tells me that they won't reset this net. They will leave it in the water and come back and get it as they did with the nets this morning. "Normally, we stretch it back out and reset it, but since the yellows are done and he wants to move to another area to catch more whites we'll get this one later," he says.

"I'll fish the white but I've got an order for a couple hundred thousand pounds of carp. We'll use a haul seine, like you did the mud shad, to catch them. Now was it mud shad or rockfish old Walter was pullin' out of that river seine?" he asks with a grin.

"No, no rockfish," I insist, "only mud shad." I ask if they will haul seine here on the Gunpowder for the carp, too.

"No, I've got a different spot I go to for them. You have to know where they hang out to catch your money's worth. Different spots for different species. I've got it pretty well figured out, and it ain't at the Pink Palomino [a strip bar] off Charles Street in Baltimore," Danny says as the men laugh.

There are cleats placed in strategic locations all over the boat. I ask Danny if he designed and rigged the boat this way.

"These are all my own inventions. I play with somethin' for a while one way or another until I figure what's gonna work best. I've been playin' at this for over thirty years and there's always a better damn way to invent a process."

As they cull the fish, Joe shows me a couple of fish that haven't spawned yet. Eggs are all over the culling board. Danny tells me that ribbons of eggs hang off the fyke nets and they make a great net hatchery. He also tells me they will survive better than the eggs on the bottom. Not bad, I think, for their own brand of aquaculture. "Sometimes when we see ribbons all over a net we'll leave it, let it spawn, and come back later to empty it." Danny says. "The state, in all their wisdom, put an eleven-inch cap on us and then took two weeks in February away from us. We try to help the fishery by leaving these ribbons of eggs and they take it away from us anyway," he says.

Joe takes a yellow perch from the culling board and squeezes snow-white sperm over the ribbons of eggs on the board. "That's artificial insemination watermen's style!" he exclaims, smiling. They do this regularly throughout the day and shove the fertilized eggs off the board into the water.

They usually fish about 102 nets. The result, for market price, is catfish, large white perch, eight-inch white perch, and yellow perch. They generally pull between 10 and 20 nets a day. Today they will pull 18. Of the original 102 there are about 60 nets still out in the river that will eventually be pulled in over the coming weeks. "Because the yellows spawned so early and we lost two weeks of the season, if we left these nets out another week we wouldn't get another hundred pounds of yellows so we'll pull 'em." Danny says. Everything is calculated for the most efficient use of men and equipment, working the resource to its full potential.

With all the nets emptied, there are tubs full of yellow perch, catfish, and large whites. The large holding box is over half full of white perch. We head back to the dock and "old man."

After we're tied up, Danny gets out of the boat and notices the fish he left earlier for his heron is gone. "Well, we stayed out longer today. He came and got that fish and probably some more from the boxes over there. He may show up in a minute though. We see him every day."

Joe takes the forklift and goes up to one of the sheds, returning with a full load of wax-coated boxes that hold fifty pounds of fish each. He sets the pallet on the dock by the boat. Then the three of us go up to the icehouse to load up twenty fifty-pound bags of ice. He drives the ice down by the boat and leaves that pallet on the forks. Danny has disappeared into the house to talk to Joyce for a minute.

Dan stays in the boat and hands Joe the culling board. With that out of the way, he takes a large dip net and sets it next to the holding box. Then as he separates the tubs of large whites, yellows, and catfish, Joe takes six of the boxes, lines them up alongside each other on the step going up to the pier, and fills them halfway up with ice from one of the bags. Once he has the boxes laid out, he gets three large plastic buckets with holes in the bottom and sets them on the ground between the boat and a set of scales hanging from the boathouse room.

Danny has returned and helps by placing the boxes on a pallet. Dan begins to scoop yellow perch from their tubs into the plastic buckets. Joe lifts one by the handle and hangs it from the scale. A small black line marked on the scale shows fifty pounds; another line shows sixty pounds. Yellow perch are put sixty pounds to the box. All the other fish make fifty-pound boxes.

Danny looks at the scale and motions for Joe to add a few more fish to the bucket. When it reaches the sixty-pound mark he says, "Okay," and Joe lifts it from the scale and pours the fish into one of the boxes on the pallet. The fish fill the box almost to the top. Danny tops it off with more ice and places the top on it. With a black wax pen he writes, "YP-60lbs."

This process continues through the yellow, large white, eight-inch white, and catfish. At the end of the day there are 40 pounds of catfish in a box, 186 pounds of large white perch, 560 pounds of yellow perch, and 806 pounds of white perch. Danny has marked each box appropriately and written his totals on a piece of cardboard box to be taken to the house and transferred to his diary and books. This day's market will bring $.40 for the white and $1.40 for the yellow. Danny is relatively pleased.

As Joe and Dan clean up the boat and tubs and make ready for the same routine tomorrow, Danny tells me that he will use a different size net with bigger holes in the net mesh for the larger white perch he will hunt soon. "Bigger mesh size and plain bigger nets will hold more fish. We also set them in deeper water where there's more tide and the bigger nets keep the fish from beatin' each other up. We'll still cull 'em, you understand, even though there's no maximum size limit on the large whites. The maximum size limit on yellow perch is eleven inches and they just put that in this year. There's no size limit for the hook-and-liners and they catch all they want even if they can only take five yellow perch. The association tries to help the recreational fishermen get more than a five-fish limit and the CCA, their buddies, shoots it down. That's because the CCA is a bunch of catch-and-releasers who don't want anyone taking fish."

I leave Danny and Joe under the boathouse at the fish cleaning station taking herring from a five-gallon bucket. Joe is slitting them up the middle, scooping the roe (eggs) out and putting it in a plastic Coke bottle with the neck cut off. Danny is cleaning herring and perch and collecting a little roe for himself. Dan is sitting on the step smoking a cigarette. Before I can thank them all Danny asks me to stay for dinner. I thank him but decline. I've been gone over fourteen hours and I hate perch and herring roe. I promise Joe and Dan I will send them copies of *Sunup to Sundown* and they are happy with the offer. Danny emphasizes that he and Joyce would like me to stay and again I sadly decline and walk slowly to my Jeep. His last statement of the day to me is, "Come back any time. I'll help you any way I can," and I recall Kenny Keen's words of encouragement and offer to help Danny at the convention in Ocean City. Now I've heard the same from this bear of a man with the angry, tough façade and the heart of gold underneath. I leave thinking he is a good man and I will return soon.

Sadly, on July 12, 2000, Danny Beck was found guilty of illegally harvesting rockfish in federal court. The Fish and Wildlife Service accused him, under the Lacey Act, of taking as much as $70,000 worth of rockfish from the waters off the army's Aberdeen Proving Ground in 1999.

While he admitted to catching fish using licenses allocated to his wife and daughter and to falsifying catch information, he insisted he did not know he was violating federal law. "I knew I was in federal waters but I didn't know it was a federal violation unless it was shipped out of state," he told U.S. District Judge Marvin J. Garbis. "No waterman knows that."

Larry Simns, president of the Maryland Watermen's Association, agreed with Beck and complained that he was "boxed in" and had no choice but to plead guilty because he couldn't afford the expense of a federal trial.

"We don't condone what Danny did. If he did something wrong, he should pay the price for it, but it should be in Maryland court and not federal. The authorities have stretched this thing a long ways to make a federal case of it," Simns said.

Beck could have faced up to five years in jail and a $250,000 fine when he was sentenced on September 26, 2000. However, in August, before the sentencing took place, a plea bargain was reached. Beck received a sentence of one year and one day (he will serve about ten months), a fine of $3,000, and the loss of his striped bass permit. To regain his permit, he will have to go on the state's waiting list. With over two hundred fishermen on the list and a turnover rate of only four or five a year, Beck will never fish for striped bass again.

After all the negative discourse about this waterman, it was time for the community to take action and express their opinions. They sent an overwhelming number of letters to the judge describing Danny Beck's dedication and contributions to his community, the watermen's association, and individual watermen. In response to the community, the judge awarded a shorter sentence in a prerelease center in Baltimore County in a work release program. He worked at Markley's Marina, saw his wife Joyce during the day, talked on the telephone, and did his time. No waterman should go to prison over a fish. Danny Beck will be back and many will be thankful.

CHAPTER 8

Moving into Spring

Moving into spring on the Chesapeake means putting oyster rigs away, except for those planting seed oysters, and refitting for crab or fishing season. Pound nets are going up for those fishing the tributaries, and crab pots are being assembled, fitted with zincs to repel algae, and stacked in yards waiting to be placed in the water, which could happen as early as the end of April. As the 1998–99 oyster season drew to a close, the preliminary harvest of 302,000 bushels showed a 70 percent increase over the 1997 harvest with a dockside value of almost $6 million. Then, to the great surprise of many, oystermen were granted a reprieve, in a still very slow season marred by bad weather, with a two-week extension of the season. Some scrambled to catch more oysters and others, like Russell Dize, a skipjack captain from Tilghman Island, signed contracts with Dorchester County to collect seed oysters and then replant them in areas where they might prosper for next year's catch.

As seed oysters are spread by watermen with the hope of creating new beds free of MSX or Dermo parasites, the Oyster Recovery Partnership is planning to take on the challenge of placing forty million baby oysters in the Bay throughout the summer and early fall. Staff and volunteers will move and plant up to 50,000 shellbags, small bags of shell that are placed in the water—quite an effort and quite a contribution.

I called Russell one evening in early April and told him I'd like to go either oystering or seeding, whichever he was doing. He said that he stopped dredging for oysters but did have a contract to haul seed oysters and that I was more than welcome to come along. My instructions were the same they had been ten years before: "Meet me at my dock a little before five o'clock Monday morning but call me Sunday evenin' to make sure we're going out."

Without waiting to call Sunday evening I planned to drive two hours to Tilghman and spend the night at Harrison's Chesapeake House, then drive a couple of blocks to RDS Seafood, Russell's place, at about 4:40 A.M. It would be far more pleasant to do that than to leave Fairhaven at 2:30 A.M. to make sure I wasn't late. I'd made too many midnight drives around the state to catch a departing boat while working on *Sunup to Sundown*. I'm a little wiser this time around. The only risk was the weather; I could easily drive over, spend the night, and waste the time if the wind picked up to heavy seas canceling the trip. Things looked good, however. The Weather Channel was calling for rain on Monday but no wind to speak of. Rain did not stop them from going out; wind did.

I made the leisurely drive early Sunday evening. As I was crossing the Chesapeake Bay Bridge, the sun was dropping behind the horizon, covering the water in a light orange shroud like a linen blanket. There were no whitecaps; the Bay was like glass as I looked over the railings from a hundred feet in the air. I came eyeball to eyeball with a gull that had snatched a wind current and was lazily hanging in suspended animation as I drove past at fifty miles an hour. I took a deep breath and thought about the quiet, relaxing drive down the Eastern Shore to Tilghman.

It was dark when I left St. Michaels and headed down the lonely, eleven-mile stretch to the island. Around dusk I am always vigilant for white-tailed deer moving across rural Maryland roads. The population continues to multiply at an incredibly healthy rate, to the aggravation of the farm community and nighttime drivers. Driving to Tilghman on a Sunday night makes you rapidly aware that the farther you get from

Easton and St. Michaels the fewer and fewer companions you find on the road. When I arrived, the island was asleep, its residents tucked away for the evening. The parking lot, restaurant, and bar at Harrison's all bespoke the end of the weekend. Perhaps half a dozen cars and pickup trucks were out front, their owners, more than likely, still in the restaurant or bar. When I checked in, I asked about a wake-up call and was told that no one would be on duty when I needed to get up. No problem, I thought until I opened the door to my room and found there was no alarm clock and no telephone. I wallowed in anxiety for a minute before I remembered that Cindy's cellular telephone was in the Jeep. I found it and called her to ask for a wake-up call at 4:00 A.M. It was no problem from her end, but I then realized that the battery was low and that I would have to leave the phone on all night to make sure the call came through. The solution was easy enough. I decided to stay up until eleven or so and then the phone would only have to be on for five hours.

The plan worked. I received the call, left my room key on the dresser, and drove from Harrison's at 4:45 Monday morning. It was sixty-five degrees and pitch black outside as clouds from the approaching front acted as a shield protecting me from any light of the morning moon. Somehow I knew it would be a good day.

When I pulled into Russell's parking lot, four or five RDS Seafood trucks were parked haphazardly along the property lines and in front of a shed. I parked beside one of them figuring I wouldn't be in anyone's way during the day. RDS is Russell's wholesale seafood business. He ships crabs supplied by local watermen all summer to places as far away as New York. I finally saw his pickup truck parked by the pier that ran parallel to the shore. Walking over, I got a good look at his old skipjack, *Kathryn*, built in 1901, resting patiently in the glow of streetlights mounted on poles by the shed. Russell was coming up from the cabin as I was stepping aboard. Daylight is still well over an hour away.

Russell Dize, fifty-nine, stands several inches under six feet. With a slight build and wearing jeans, chamois shirt, wire-rimmed glasses, and neat haircut, he looks like the hardware store owner and operator that he is, hardly the stereotypical picture of a seasoned waterman. A quiet, soft-spoken man, he is genuinely glad to see me and eager to help out with my project.

"Put your bag below and come on back up. I'm waiting for my crew. I've got four head comin' from Cambridge. One boy's worked for me but the other three are inexperienced and have never worked a boat before. This is goin' to be a very interesting day."

As I go below, I see a small stove directly in front of me in the cabin. There are shelves on either side holding salt, pepper, other spices, bread, luncheon meat, and breakfast sweet rolls. A drawer has silverware and other utensils. Bench seats are positioned on either side below the shelves, and behind them, under the deck, are two berths. There are more berths forward. The cabin is about six feet by six feet, cramped quarters for five big men, but warm and dry. Water for coffee is on the stove, ready to be heated. Sodas are on deck. I shove my duffel bag under the starboard side of the cabin and sit on the bench looking around. On a beam to the right of the stove is a plaque with Psalm 107: 29–30. I think for a moment about Russell's upbringing on Smith Island where a strong Methodist faith helps maintain an optimistic outlook offshore.

The musty smell of damp oilskins and oyster shells permeates every inch of the cabin and I am reminded of how warm and stuffy the cabin can get, aided by the smell of perspiration and wet rubber, when it is snowing and well below freezing. It has been too long since I've been belowdecks aboard *Kathryn*.

Putting my head up through the hatchway I take a breath of spring air and gaze into the darkness at a few lights, which look like candles, in the windows of those rising early. It is quiet except for the sounds of Russell shuffling things around behind the wheel and the three ducks talking to each other as they swim back and forth between the pier and shore, visible only when they come from the shadows into the dim light from the shed.

While we wait for the crew to arrive I ask Russell how his children are and what they are doing. Russell didn't let his son Rusty work on the water as he was growing up, although he did let him work in his hardware store and later drive a truck for RDS Seafood. "They're fine. Rusty is a lawyer working in Connecticut and Leah is an overeducated dental hygienist working on Capitol Hill. Rusty just got married to a fine young girl who has a good job with Xerox. I see them because they come home pretty regular but I'd be lying if I said I didn't miss them terribly. They're doing what I wanted them to do: get an education and a strong job off the water. I guess we never prepared ourselves for the fact that they would probably leave the island and the Shore."

Hauling seed oysters is vital to the process of transplanting juvenile oysters from beds where they are thriving to depleted beds, where it is hoped they will continue to grow in spite of pressures from parasites and disease. It is an important program, funded by the state or, in this case, by Dorchester County, which pays watermen to dredge and transplant the spat, which is attached to shells. Today, Russell would like to make two runs, if the weather holds, dredging two thousand bushels with each load. The skipjack will carry them the 13.5 miles to the dump site, and the three skipjacks working today will each get $1 per bushel. Their contracts call for five thousand bushels each to meet the requirements for a $15,000 county grant provided by the oyster seed program.

Russell moves below to look through his log to find the location of the seed oysters. "I usually write down loran coordinates and can follow them in the dark. When I need them I can't find them so now we're going to have to wait until we have some light." I realize later that we needed the light to see the red flags marking the bed.

"We'll use the yawl to push this ninety-eight-year-old lady to the seed bed. We'll dredge 'em up and haul 'em to where the county wants us to dump them," he explains. The yawl, or push boat, is a small powerboat with a four-foot beam. It has enough room to carry a large inboard engine and hangs off the stern of the skipjack by an intricate array of lines and pulleys. In the late 1960s, laws were passed permitting skipjack captains to dredge under power on Mondays and Tuesdays, while still sail dredging Wednesday through Friday. The yawls are used for the power since skipjacks are not fitted with engines.

"I swore, when I was young on Smith Island and workin' with my father, that I would never work a skipjack after growing up on them. Then

Captain Russell Dize aboard the skipjack *Kathryn* talks on a cell phone with another waterman. This "new" technology allows men on the water to call their families, their dealers, or their distributors.

I realized how much I love to run her under sail and I wouldn't trade it for anything else in the world. I don't need to do this anymore but I still love it. My wife is jealous of this old boat but I swear when I stop sailing her I'll turn her over to a museum or school. I'll never sell her.

"Dredging seed oysters isn't rocket science and doesn't take as much figurin' as dredging for oysters off a good bed but it's good solid money before I start getting my crab business going." He tells me he will buy from as many as thirty different crabbers every day and deliver them in three or four trucks, depending on which ones are running.

As we talk, four large black men come from the darkness and are suddenly on the pier walking to the boat. Russell greets his man,

Philip, who has brought three other men with him. We later learn one is Philip's uncle. Philip is the only experienced one. The others have never been on the water. He explains the power dredge, which looks like a bear trap with a four-foot opening and a rope basket attached to it. When the bucket is dropped overboard, the open mouth is dragged along the bottom swallowing up oysters, or in this case, shells with seed oysters attached to them. When the captain feels the dredge pull he can tell when the four-foot-square basket is full and will bring it up with a hydraulic winch. The haul-in winch controls are located by the wheel. Another power winch is located amidships, where the widest part of the deck can stretch an easy twenty feet on a skipjack. This one is operated by one of the men and lifts the full bucket up over the edge of the rail so that the other two men can each grab hold of a rope handle, one on either side, and haul it over the rail to dump the shells on the deck. It actually works in one smooth motion with Russell pulling in the dredge, a crewmember taking it to the edge of the rail, and two men dumping it on board. It is then dispatched back over the side, making sure the rope bucket is not twisted, and the process will be repeated as soon as Russell feels the tug on the line telling him that it's full. He tells two men to work each side of the boat and gives them instructions on hauling, dumping, and shoveling. I stand back by the wheel and listen. When he notices that the men are not equipped for working the water, he offers them oilskins.

Each man takes a pair of oilskins and they all begin to laugh and joke. "I ain't never been on the water. This is my first time and I can't figure out how to put on these damn pants," one says. As the others laugh, Philip helps him with his straps. "I ain't no waterman. I ain't into no playin' on the water, either," he says loud enough to wake the rest of Tilghman Island as the others joke and laugh at him. He calls Philip "Tree" because he stands at least six feet, four inches tall. "Hey, Tree," he yells to Philip who has moved back near Russell and me, "make sure the captain there keeps that radio workin' 'cause I ain't gonna get stuck out there in the middle of no bay!"

A dredge bucket sits idle on the deck of *Kathryn* before being sent over-board for more oysters.

Even I am smiling at this man's nonstop banter. I can't tell if he's talking because he's nervous or because it's his natural tendency. I find out soon enough that it is a little of both.

Russell tells Philip to get a couple of the men to come back and help lower the push boat into position directly behind *Kathryn*. As it is hand lowered by different lines and pulleys, Russell is calling out orders and instructing them to go slow and to not get hurt. I can tell he is a little nervous about what he's gotten into with this makeshift crew of inexperienced men. With the boat in place he tells Philip to untie the skipjack as he starts the push boat's engine. The quiet, peaceful hum bounces off the night and echoes across the harbor as a clam boat, with its low sounding diesel, goes slowly by.

"Bust their asses, Russell," the captain of the clam boat yells.

Russell laughs as our new comedian, "Eddie Murphy," as we call him for the moment, says, "No, I ain't goin' out on any small boat like that [the forty-two-foot clammer]!"

The three resident ducks paddle alongside to see what all the commotion is about. "Eddie Murphy" looks at Russell and says, "Are they part of your family? 'Cause if they ain't, I can help weed 'em out for ya."

When we're ready to pull away Russell and I slip below, so I can heat water for coffee and he can put on his oilskins and rubber boots. As we're going back up, he turns to me and says more seriously than I've ever seen him, "This is goin' to be something today. I hope nobody gets hurt."

Once he's behind the wheel he tells the men they can go below and get coffee and sweet buns, reminding them on the way that "all I want you guys to do is shovel. We'll do the rest. Haul and shovel, that's it." He turns to me as I look forward and says again as he slowly shakes his head, "This is goin' to be some day." I decide to go back down below as we leave for the channel marker as the slightest hint of dawn breaks the horizon.

Philip's uncle sits on one of the bunks pondering the day ahead of him. "I ain't never been on the water before like this. I love to fish now, so don't get me wrong. I only missed four days of fishing all last season. It's peaceful. I'm a happy person. I make it through every day with a smile and happy feeling. No prejudice. That stuff will eat you up inside. I don't use no words like 'racial' and all. No, sir. We just need a little more love is all."

"Well, you tell me in seven or eight hours if you love oysters!" I say.

A wide, toothy grin stretches across his face. He takes a deep breath, swelling a barrel-sized chest with arms to match. I realize that his strength will not carry him through the day shoveling oyster shells. He will need to develop finesse and rhythm or all the muscles in the world won't save his back.

With a cup of coffee in my hand I go up on deck to talk with Russell. He's on the radio. "If it's blowing fifteen to twenty out of the southwest it means we're not runnin' too far. Let me know what it's like when you get out of the river.

"With the weight of all these oysters on board we don't want to get caught in a breeze," he tells me.

He pauses to listen to the radio and we smile as a man announces that he just got new oilskins but "they're so big I look like an orange clown."

Russell leaves the wheel and moves up where the power dredge sits amidships. It will lower and raise the dredge buckets as instructed by the man at the controls. "If I'd found my notes and hand coordinates we'd be dredgin' already. Instead we've got to look for flags that mark the bed where the seed is." The men lay dredge buckets on either side of the boat with the rope bucket hanging over the side. They will work two men to a bucket.

He tells them to begin putting up plywood barricades along both sides, fore and aft, except for a six-foot area where the dredge buckets are operated. As shells are emptied on deck the men will shovel them back as far as the wheel and up to the bow against the four-foot-high plywood.

In the dark I see the lights of two other skipjacks: Bart Murphy on *Nellie Byrd* and Lawrence Murphy on *Thomas Clyde*. Both men are experienced skipjack captains and we will work the bed and drop seed together for the remainder of the day. Once we've arrived Russell tells Philip to let the starboard dredge go over the side. Russell looks over the side and watches the line. As the teeth on the bucket grab into the bed of shells the line pulls tight and it looks like we've caught a huge fish.

Three minutes later Russell tells the other men to let the port side dredge go over. Now the two lines look like wings fanning out on either side of the boat. After a few minutes Russell can tell by the feel, by the pull on the line, if a bucket is relatively full. As he raises a bucket from controls by the wheel, he tells me to take the wheel. He goes up front to show the new men how to work the winch that will lift the bucket its final foot up over the side so that they can each grab a rope handle and lift it over, emptying as they move.

As the men shovel seed oyster shells aft we are just off the stern of *Nellie Byrd*. The boats are working the bed in clockwise fashion. Russell yells to have them raise the buckets again and for the men on the port side to shovel aft and the men on the starboard side to shovel forward.

He yells that they must keep the shells balanced as they shovel in the four corners created by the plywood cage. He watches and I can tell that he is already both frustrated and anxious.

"Take the wheel and slow her down. I've got to go up front," he tells me as he climbs over the plywood wall closest to the wheel and hustles toward the men telling them, "Get those dredges up quick and don't pull them too far over on the deck. When they've reached the topside pull 'em over, empty 'em, and let 'em go again. Come on, boys, you've got to keep things movin'. If you don't empty 'em right you'll kill yourselves before we even get started. Now come on, find a rhythm and get into it."

For the next fifteen minutes he hops all around the deck checking the dredges, showing them when and how to empty them in the easiest and most efficient way. I'm still at the wheel keeping the boat on course.

It is growing cloudy and overcast as daylight comes. Russell comes back to the wheel and takes over. "I could be using tongers for this work but since the state extended the season for them by two weeks I've got to make do. Other boys are already making ready for crabbing. Dorchester County is paying a dollar a bushel and we'll take five thousand bushels over three days if these boys show up after today."

Observing the men unload the buckets and shovel he says, "The rhythm isn't in the boy who smiles all the time. His back is hurt already and we've just started. They aren't letting the dredge empty itself with a little help from them. They're forcing it." He tells Philip to change places with "Smiley" and to work the starboard side.

Half an hour later Russell is frustrated and it's showing in his tone and body language. "He just can't get it. He was smiling earlier but he's about to cry now!

"My regular crew was legendary. Before they finished shoveling I had a bucket in their face ready to go again. They were unbelievable. When I'm workin' it hard I have two men on each dredge and one man workin' the power winch for both sides. We've got 'em coming fast. I'm not workin' it today. We're moving slow because they can't seem to work in unison here. Now, see that one boy there, he's already hurt. I'm wondering if he'll be back tomorrow. If he doesn't show I don't know who I'll get to work. That's the hardest part about this business these days."

I decide to try to get him to shift his own gears for a minute to ease his tension. I ask what he thinks of the state of the Bay as we enter the new millennium.

"We have a healthier Bay now than I've seen in some time. Don't get me wrong. It could be better, but if we didn't have MSX we'd be up to our tailbones in oysters. And rockfish, there's more than I've seen since I was a boy. Now, rock aren't too good for crabs because they love to eat 'em but there's still right many fish out there." He hands me the wheel and climbs up front again.

Minutes later he returns. "You know, I don't have to do this anymore. With RDS Seafood I can make it fine. Why, I swore I'd never do this in the first place. I've been workin' skipjacks all my life and I've had *Kathryn* twenty years this year. My wife says I like her more than anyone else. That's not entirely true," he says with a smile. "She's the best in her own right—my wife, I mean.

"Lawrence is gettin' three licks to our two. It was the other way around before, but he's still got a dredgin' crew and he's usin' a different dredge. He's using a hard bottom dredge. I'll change mine after this load.

"Yeah, finding crew is the hard part. I could have caught a lot of oysters this year under sail, which is what I love, but boys make six or eight hundred dollars in two days and they want to go home. Forget about finishing the week." Many of the captains find this to be true around the Bay. Young men can make good, quick money working the water but they seldom see the benefit of saving or investing. As Russell says, "They look for instant gratification instead of the future. That's being young and invincible, I guess."

As we enter our second hour, the boat is loaded with seed oysters up to the top of the plywood all around the boat. She is sitting low in the water under tremendous weight and I understand why Russell and the other captains want to watch the weather. Bart Murphy, with a full load of seed, calls on the radio and asks Russell where he wants to go to

Captain Russell Dize stands on a pile of seed oysters five feet high, equal in height to the roof of the cabin.

unload. Russell says they will head for the Little Choptank River today and that Dorchester County wants them dropped in a different location tomorrow. Lawrence is on the radio saying he has a funeral to go to on Wednesday so he'd like the other captains "to make two runs today and tomorrow and make Wednesday the shortest of the three." Bart and Russell say they're flexible. The men want to run together because Russell has the loran coordinates for the location where they will dredge seed Tuesday and Wednesday. They also like working the water in close proximity to other men in case something goes wrong on the skipjack. They didn't need the coordinates today because they knew about where to go off Holland Point and could see the flags.

The other skipjacks leave the bed as Russell instructs his crew to catch a few more. He hands me the wheel and climbs over the plywood, now teeming with shells, like a kid climbing over a sand dune and just about as fast. "I ain't goin' down without a good load, boys. So let's go. Bart says he's ten minutes ahead of us now and I don't want them too far out there."

Within five minutes he says, "Bring in the dredges and relax for a while. Go get some coffee, rolls, soda, whatever you want." I move below to make sure there is hot water for coffee and that the sweet rolls are out. I figure it's the least I can do and it helps them all a little. Sitting on the bunks, out of their oilskins, the men can reminisce about their first experience as watermen.

"I feel pretty good," says the biggest and clearly the most effective of the four. "I'm an ex-football player. When you tell me it's the fourth quarter, you can bet I'll be ready!"

Philip's uncle, "Smiley," is sore. "I ain't never gonna say nothin' 'bout no waterman workin' on the water. No, sir. I ain't never gonna say nary a word. It looks easy but by God I wasn't sure I was goin' to make it!" The other men smile at him.

"You think this is hard," says the football player. "You ought to pick watermelon. Now that will kill you!" After a few minutes, one by one, the men get silent and fall into their own worlds of rest or sleep in the cramped quarters of hidden bunks under the deck or benches by the stove.

After the seed oysters are dredged, they are transported to another place on the river to be unloaded. Here, the pickup crew of *Kathryn* shovels shells overboard.

Smiley closes the conversation by saying, "I ain't goin' up under there in no dark bunk. I get claustrophobia and besides I don't know what's up under there sleepin' with me!" I smile at him and go topside to talk to Russell. It will take some time to travel some thirteen miles up the river to the dumpsite.

"Watching those men this morning caused me some worry," Russell tells me. "It would be bad if one of 'em got hurt. Workman's comp cost me fifty cents on the dollar out here and five cents on the dollar in my hardware store. It's a mess and the state's tryin' to help, but right now that's prohibitive."

At 9:55 A.M. as we move up the river, Russell travels into his own thoughts and I move into mine. Without realizing it I find myself meditating on rain hitting the water. I put up my slicker hood and with the very mild temperatures I am more comfortable out here than I have been in some time. I recall a time when I was ten or eleven and camping with my friend Bob Burack. We were in a small pup tent with no door flaps. I woke at the break of dawn to watch rain falling straight down on our camp. I was dry and warm in my sleeping bag—that was the only other time in my life I remember being so comfortable and peaceful. The rain continues to pelt the water in a mesmerizing rhythm.

Off to our starboard side we pass the skipjack, *City of Crisfield,* dredging under sail. Farther up the river as we near the dumpsite, we pass three patent tongers working beds close to shore and taking advantage of two extra weeks of oyster money. Marvin Gaye's voice echoes across the water between raindrops. The men below are singing in unison, something they need to do with a shovel in their hands, Russell notes. "We get two loads today, two tomorrow, and one Wednesday, we'll get $5,000 a boat. It's guaranteed money, so it's good after a season of worrying about who's working when and if we'll catch our quota."

I take the wheel for the final half hour as Russell goes up to the bow with the men who have now come up on deck refreshed and ready to shovel a thousand bushels of seed overboard. As we enter the dredge dump site, marked by flags attached to floating buoys, a marine police officer out of St. Michaels, Maryland, comes over and stays with us for the thirty minutes it will take the men to shovel the oysters overboard. Russell knows him and says, "They don't bother us when we're workin' seed as long as we stay within the buoys, but we get hounded sometimes when we're dredgin' oysters. When we're cullin' sometimes they get mixed up and a small one will get by. That's illegal and I'll get a nice ticket. You can almost expect it when you're workin' with an inexperienced crew." I climb over the plywood and oyster shells to help shovel overboard the seed that is more than a foot deep and has covered the hatch and cabin. "It's hard to shovel in those slickers. Take 'em off and it will be easier on you," he instructs. We all shed our coats and accept a few raindrops gladly as sweat begins to run from our brows. It doesn't take long before my back begins to hurt but, for some strange reason, I feel compelled to stay the course and not weaken under the strain. I am not one of them, but for a few minutes I can endure and pretend. It is worth the pain.

When the final shell is unloaded, the plywood blockades are stored along the bulkhead, and the deck is washed clean of shells and mud, we can relax for the long trip back to the dock. The men strip off their oilskin overalls and boots and head below to lie in bunks or sit quietly. They are considerably more reserved and they keep their own thoughts—much more so than this morning when they were excited and apprehensive like children waiting to go on a ride at the carnival. I believe they found this day with Russell as interesting as I did. While their perspective may be far different—I will never know—I can't help but think that they were intrigued by the experience. And, like me, they will not make a career of the water but they have gained new respect for the men who do.

Russell and I are also more sedate. We talk about how fortunate he is that no one got hurt, his children living away from the island, the movie *The Sixth Sense* that he encourages me to see, the Watermen's Association, and the state of the Bay.

After a while Russell wants to go up to the bow and asks me to take the helm again. We are still the third skipjack in the parade, and I am enjoying my turn at the wheel. Over the course of the next hour or so, Russell will turn the wheel over to me two or three times, and I am pleased with his confidence in me.

"Tree" is a big man who had never before worked the water. He worked all day without tiring. Here, he settles in the cabin of *Kathryn* for the two-hour trip back to the dock.

When we reach the harbor Russell tells the men to come topside. He needs to raise the yawl and tie off the boat once we're at the dock. Even with the pulleys it's still difficult and it takes all available hands to get the jobs done. When we are secure at the dock, the crewmembers gather their belongings and Russell tries to make sure each will return for the same routine the next day. "I think they'll all be back except for the one with the sore back. He may show but we'll have to see. If none of them show I'm in a fix so tonight I'll call Philip to double-check. With boys getting ready to crab and others still tonging my crew selection will be thin."

We collect our things and walk off the boat together. I thank him for a good day and he lets me know I am welcome anytime. I assure him that I will call.

On the two-hour ride home I have time to think about working with these men on the water. Their lives are formed day to day—immediate success or failure—and each day presents its own trials and tribulations. Rarely are they uneventful. It is certain that when I am working with them on the water, Washington, politics, world affairs, and day-to-day anxieties are irrelevant. They have no consequence in the middle of a river or the Bay. I am tired, relaxed, and full of thoughts of shoveling seed oysters and steering the skipjack. Things are good.

CHAPTER 9

Bay-Related Legislation

The oyster season ended on April 9, only two days before the April 11 adjournment of the Maryland General Assembly. Virginia's assembly had recessed weeks before, and with both sessions over, watermen were relieved to have faced few legislative and regulatory issues affecting their livelihoods. Instead, the major issue addressed during the session was the dredging of Baltimore Harbor and the dumping of the spoil in part of Site 104, a deep trough north of the Chesapeake Bay Bridge off the shore of Kent Island, where some past dumping has occurred. Another deeper trough, located just south of the bridge, is a place watermen do not want any dumping ever to occur.

Watermen have always opposed dumping dredge material in the southern trough, perhaps the deepest water in the Bay, because it is known as an invaluable spawning ground and wintering area for a variety of fish. It also happens to be a valuable fishing area for both commercial and recreational fishermen. To dump spoils in the deep trough would create a huge negative impact on its ecosystem. The government has finally agreed not to put any dredge spoils in the area.

The placement of dredge material at Site 104 was a contentious issue in which the state of Maryland and the U.S. Army Corps of Engineers proposed to use the four-mile-long, one-hundred-foot-deep area for spoils from the shipping channel in Baltimore Harbor. It was also a controversial issue with watermen, members of Congress, environmentalists, local governments, and area citizens. Newspapers across the state gave daily updates and dramatic descriptions of the latest arguments. The Maryland Watermen's Association unexpectedly contradicted tradition and some of its members by speaking in favor of the dumping at Site 104 in exchange for millions of dollars for oyster replenishment.

At the end of the legislative session, all bills relating to Site 104 died as they had in previous years. The corps released its revised draft of the Environmental Impact Statement in 2000 with a final decision in 2001 not to propose using Site 104 for dumping dredge spoils.

The following legislation passed during the 2000 session:

- HB 407 states that upon the death of a licensee an authorized representative can indicate the permanent transfer of the tidewater fishing license to an approved relative.
- HB 1118 allows for a freshwater fishing guide license for nontidal waters after completing an application and paying the fee. It also limits the number of freshwater fishing guides to 150 annually and establishes a waiting list for those over that number.
- HB 1254 establishes a DNR program to study the impact of recreational watercraft on submerged aquatic vegetation (SAV) beds.
- SB 332 gives the DNR authority to adopt regulations on the placement of pound or stake nets. One example is the establishment of a limit of eight assigned locations for net placement per licensee.
- SB 639 allows for the creation of a state debt for the development of the watermen's monument at Kent Narrows in Queen Anne's County. The Queen Anne's County Watermen's Association must provide matching grants.

Thirteen bills relating to watermen's interest did not pass. Many of these bills would have added regulations making it even more difficult for watermen to earn a living. None of the bills pertaining to open water dredge material placement passed.

"The legislature was good to us this year," said Larry Simns, president of the MWA. "There weren't any harsh measures that would hamper watermen from making a living. We've got to be alert every minute from January to April every year. The Chesapeake Bay Commission, for example, proposed a bill that would have further restricted clammers from working in SAV beds. While we originally opposed the bill, we sat down and worked out our differences with the commission. We put amendments on the bill that would create a 'no loss, no gain' of clamming bottom so this was a good success 'cause we also worked out straightening the SAV lines so that it will be easier for clammers to work them and easier for police to patrol them. Also, fewer buoys would have to be placed. We think this is all in the right direction and all sides are winning in the compromise. It ain't never easy, you know, but we usually

Watermen congregate outside the State House waiting for new crab regulation discussions to begin. Courtesy Maryland Watermen's Association.

come out all right. The legislators are pretty good to us. I'd rather be out fishing but if you turn your back for one minute they'll mess up everything you been workin' on," he says, and I know that without his working the halls and committees, the watermen probably would have faced harsh bills that would have impacted the way they make a living.

If the watermen made out well in the state legislature they did equally as well in Congress. A bill was overwhelmingly approved by Congress to extend the federal-state Bay cleanup partnership for six more years and to increase funding for the Chesapeake restoration effort. The Chesapeake Bay Restoration Act authorizes the continued operation of the Bay Program through 2005 and calls for giving it up to $30 million a year, nearly $12 million more than it receives now.

In addition, the bill requires federal agencies to adhere to the Bay Program's nutrient reduction, habitat restoration, and other goals when managing land in the watershed. The Bay Program is a partnership that includes Maryland, Pennsylvania, Virginia, the District of Columbia, and the federal government by way of the EPA. More than twenty-five federal agencies are involved in the Bay Program activities.

Also in May, the House of Representatives approved the Conservation and Reinvestment Act that could prove helpful in meeting the Chesapeake 2000 Agreement's goal for 2010—to permanently protect one-fifth of the watershed, ensuring it will remain undeveloped. This agreement—among Pennsylvania, Maryland, Virginia, and Washington, D.C.—is intended to improve the environmental quality of the ecosystem and ensure that less land is used for development. The legislation would also make money available to wildlife and fisheries programs, allowing scientists to gather more information about the health of species that have historically received little attention.

Overall, per year, the bill would provide Pennsylvania with $50 million, Maryland with $37 million, Virginia with $51 million, and the District of Columbia with $7 million over a fifteen-year period—a total of $2.2 billion.

CHAPTER 10

Threats to Bay Fisheries

On May 14, 2000, the federal government banned large gill nets, those with mesh openings of six inches or larger, because of the abnormally high number of dead loggerhead turtles that washed ashore on the North Carolina coast—280 of them. Portions of the lower Chesapeake Bay in Virginia were affected by the ban. The officials from the National Marine Fisheries Service suspect that nets for monkfish and possibly dogfish were to blame. The thirty-day closure will help protect sea turtles migrating into warmer waters of Virginia and Maryland. Watermen had until May 13 to retrieve prohibited gill nets left in the water.

Jerry Schill, the tenacious president of the North Carolina Fisheries Association, said that watermen will not be adversely affected by the thirty-day ban because only a few men had been fishing with the large mesh nets in recent weeks.

The nearly endangered loggerheads are being protected in the southern region, as are terrapins in the whole Bay. "Diamondbacks are not endangered, but we are concerned about them," said John Surrick of the Maryland DNR. A major reason for the concern is the disappearance of sandy beaches, the prime nesting habitat for diamondback terrapins. Waterfront property owners who have impacted watermen for the past several decades are also affecting the terrapins by building bulk-

While much publicity is generated in the Bay community about the decline of the oyster and replenishment efforts, it is important to note that other species, like the Maryland terrapin, also play a major role in the watershed's ecosystem. The Terrapin Restoration Program is an important effort to restore and preserve a species that represents another barometer to measure how the Bay is doing ecologically. Photographed by Richard Tomlinson; courtesy office of Governor Parris Glendening.

heads or revetments to keep the shoreline from eroding. These barriers also tend to eliminate the sandy beaches where terrapins lay their eggs. Maryland officials are tracking more than three hundred tagged and released diamondbacks that were captured in the wild as well as many that were incubated and raised in captivity.

As shown by the oyster recovery program and the aquaculture effort, man can reverse destructive elements in the watershed. An Annapolis environmental construction firm, Keith Underwood and Associates, has developed a system to save the terrapins. The company constructs small islands of rock and marsh grass just offshore, which slow the Bay's waves before they hit the beach, reducing their effect and limiting erosion of the beaches. "It costs about $15,000, or about the same as revetment," according to Faye Strittmatter of Edgewater-based Arundel Marine Construction. If the shoreline is not too badly eroded, the company will remove the revetments and build an island to create a new beach habitat for baby terrapins. This little bit of ingenuity can go a long way toward saving the diamondbacks.

In early 2000, an effort to protect horseshoe crabs from being overfished along the Atlantic got, well, downright crabby. Virginia refused to lower its catch under the Atlantic States Marine Fisheries Commission plan that ordered states from Florida to Maine to devise methods by May 1, 2000, to cut their catch and to compile data needed to evaluate the health of the horseshoe crab population.

Following a domino effect of sorts, Virginia's refusal to lower the catch by 25 percent could impact the conservation efforts in place by Maryland, New Jersey, Delaware, and other coastal states. Virginia has the largest harvest of horseshoe crabs and does not feel the crab is threatened.

While the harmless horseshoe crab is inedible to humans, it is the most effective bait for conch and various fish. In addition, migrating birds feed on the eggs. The horseshoe crab has survived for over 250 million years, partly because it can go a year without eating and endure extreme temperatures and salinity. Under the proposed reduction plan, landings coastwide would drop to about 1.8 million crabs. Virginia's share of that number would require it to lower its catch to 152,000 in 2000, but Virginia conch fishermen wanted their catch to be 710,000 horseshoes, and the Virginia Marine Resources Commission granted it. The U.S. Department of Commerce could have shut down the fishery completely, the Virginia legislature could have passed a bill in 2001 to change the law to allow the marine resources officials to lower the catch quota, or Virginia fishermen could have remained firm in their position and continued to argue their case. However, it never went to these extremes because Virginia came into compliance.

ALGAE IN ITS MOST DESTRUCTIVE FORMS

Harmful algae blooms, or HABs, have been the subject of considerable attention around the Chesapeake region since the pfiesteria outbreaks in 1997. HABs were relatively unknown until then and pfiesteria certainly reminded the region about the dangers and mysteries surrounding them.

Algae, microscopic plants that float in the water, are the base of the aquatic food web, capturing the sun's energy and sustaining the microscopic animals, shellfish, and fish that feed directly upon the algae or those that feed on the grazers of these plants. In some cases certain species of algae produce toxins, like pfiesteria, that impact the organisms ingesting them, or other animals in the food web, including man. It is the algae blooms with highly toxic plants that are considered HABs. Pfiesteria sounds like something out of a science fiction movie of the 1950s: The fish that are killed by the toxin are subsequently fed upon by the algae to give them nutrition.

In addition to the production of toxins, similar to what is happening in the Bay, HABs can also disrupt the larger ecosystem by being present in such overwhelming abundance as to interfere with filter-feeding shellfish or to prevent submerged aquatic vegetation from receiving sufficient light to grow. When the algae die, the decomposition process depletes water of oxygen, resulting in fish kills and what watermen call dead water.

One day in early May, I stood on the bank of our community and noticed a reddish tint to the water. I didn't think it could be a red tide, caused by an algae bloom, and my scientifically uneducated family agreed. For days the tide would creep into discussions we had while we were walking the dogs. In fact, this was one of the most dense and extensive algae blooms seen in the Bay in the past twenty years. It could not quite be considered a *red* tide—perhaps *mahogany*. The species causing the bloom, *Prorocentrum minimum* or *P. minimum,* is common in the Bay in early spring. This year's bloom spread from the Chesapeake Bay Bridge all the way to the Patuxent and lower Potomac Rivers and lasted for three weeks. Scientists claim this bloom resulted from heavy rains washing nutrients into the water, followed by unusually sunny days and warm temperatures. Fortunately for the Chesapeake Bay, harmful blooms are rare—but worth watching.

In the decade of the nineties there was much turmoil caused by a microorganism called *Pfiesteria piscicida,* belonging to a group of algae known as dinoflagellates. It has grown naturally for thousands of years in the Chesapeake in at least twenty-four life stages, of which only four are harmful. It's these four toxic life stages that caused pfiesteria in the Bay to reach headlines nationwide. The pfiesteria cells emitted toxins that stunned fish, dissolving their scales and skin, enabling the plants to feed on the fish and to cause deep lesions. Pfiesteria has been linked to widespread fish kills and fish lesions from Maryland to North Carolina. While the causes are not fully understood, scientists believe that the presence of a large number of fish in warm, brackish, poorly flushed water with high levels of nitrogen and phosphorus trigger the shift to the toxic form. Lesions and kills were dramatic in 1996, 1997, and 1998 on Maryland's lower Eastern Shore and drew the attention of legislators and scientists. Rivers were closed to fishing and swimming as watermen and recreational anglers reported seeing lesions on their hands and arms.

Since that time, the excitement caused by pfiesteria has died down, and there have been few reported kills due to the parasite. Watermen seem to be at the mercy of such invasions, just as they were with Dermo and MSX, parasites that nearly obliterated the Bay's oyster population. There is little they can do except try to work around the parasite epidemics and the impact they have on the watermen's sources of revenue.

CHAPTER 11

Crab Season

As crab season gets underway in May of 2000, when the waters of the lower Chesapeake are warming and more crabs are moving north, Maryland begins to initiate its crab marketing program in the Department of Agriculture. Plans are set for advertising and publishing news releases in newspapers throughout the mid-Atlantic region. A Web site has been developed and point-of-sale material has been distributed to wholesalers, retailers, and processors. Maryland's Seafood Marketing Program is an integral part of the national blue crab marketing initiative with the National Blue Crab Industry Association. Their logo is displayed on all containers of blue crab meat originating in the United States.

With a warm late spring sun and temperatures consistently in the eighty-degree range, it seems that crabs are on everyone's mind. Newspapers, government agencies, and the watermen speculate on this year's harvest, and the hype starts in Bay country. Crabs adorn countless hats, billboards, and logos—if "Virginia Is for Lovers," then "Maryland Is for Crabs." Virginia and Maryland watermen will harvest approximately two hundred million crabs, representing the Bay's leading moneymaker: $21.3 million in 1999 in direct sales in Virginia, for example, and much more when one considers the ancillary businesses involved, like advertisers and vendors of crab pots and related gear, cooking equipment, bushel baskets, and fuel. Virginia watermen for the most part concentrate on female crabs while Maryland crabbers focus on the elusive males. The season runs from April 1 to November 30.

Some watermen will use the traditional crab pots made of chicken wire while others will trotline. Either way, each year they hope that the harvest will be the panacea to paying off winter debts and spring boatyard bills.

CRABBING WITH STEVE SMITH

My neighbor, Steve Smith, makes his living crabbing and leaves oystering, fishing, and clamming to others. Even though it's early in the season and the crabs have not started moving in great numbers, I decided to call him to ask if I could go out with him for the day. When we connected a couple of days later he said that he would pick me up at 4:50 A.M. the following morning.

When I walk out on my porch to wait for Steve's headlights to pierce the darkness, the predawn dampness predicts a misty morning and a potentially overcast day. It's about seventy degrees and it feels comfortable. I packed a lunch—unlike the skipjack captains, other watermen take their own food and so does the crew—and I am looking forward to another day on the water.

Steve Smith, fifty-two, grew up in Ames, Iowa, but spent his summers in Fairhaven while his father, a college professor, taught at the University of Maryland. Steve had a small boat in Ames that his father trailered to southern Anne Arundel County each summer. After graduation from college, Steve and his family returned to Fairhaven full-time. He married a local girl, Barbara Beck, and bought the house where they live today, next door to his brother Scottie and up the street from his father. Both Steve and Scottie became watermen and have made a successful living since 1972 without having to do anything else. However, Scottie left the water two years ago to work with computers. Steve bought 150 pots from him and will start the season this year off Fairhaven in Herring Bay with 450 pots.

We drive a mile down the road to the pier where Steve keeps his boat. He tells me the boatyard has a new landlord: "A pretty good guy but he doesn't like the lights and noise we make in the morning. I'll turn the truck lights off when we round the bend and we won't talk once we get out. You just get on the boat and I'll load her up with gas and stuff using a small light. Then we'll sneak out and can talk once we get away from the yard."

I remain quiet as he loads baskets, clams for bait, and a gas can onto the deck, then unties the boat, starts the engine, and moves toward the channel. Noise carries across the water, and as we pass the last boat slip we hear a radio in the darkness and men, already on the water, chattering like songbirds. "That's Kenny's [Watts] radio. He forgot to turn it off yesterday. We do that sometimes. We'll get in around the same time and stand around talkin' and so forth and somebody will forget their radio's on. You know it now though. Loud, isn't it?"

During the off-season, crab pots are typically stacked in yards from one end of the Bay to the other.

Kenny Watts is another neighbor who crabs in the summer and works as a merchant mariner in the late fall and winter. He starts later in the morning on a bigger boat than Steve's twenty-eight-footer and will crab across the Bay around Poplar Island. "That's his preference," Steve says.

While Bay residents are waiting impatiently for crabs to hit the restaurants and dock bars, watermen are employing their own strategies to try to outsmart the crabs and begin the season with a rush. In truth, I've never met a watermen yet who could outsmart or figure out a crab. They can do everything right, attend to every detail, and that crab pot can be just as empty as when they tossed it overboard. Sometimes you can't figure them out no matter what you do.

Leaving the boatyard, we see Herrington Harbor North Marina a couple hundred yards dead ahead. Steve points to a small sailboat moored off the marina and silhouetted in the moon glow; it shows him which way the wind is blowing. "See, she's blowing from the southeast this morning so I'll take the channel this way."

As we leave Deale Harbor and head out into Herring Bay, we first run parallel to Fairhaven, where my house and those of our neighbors remain bathed in darkness, and then move a quarter of a mile off the Deale shore to the Bay to run his pots. Steve notices that I am wearing a T-shirt and tells me there are jackets stuffed in the beams above my head. I tell him I'm fine but can tell that he's anxious for me to be comfortable. Finally, I concede and reach up for a flannel shirt. The shirt does feel good as the breeze from the open Bay blows across the bow.

The boat is a twenty-six-footer with a nine-foot beam. It's a fiberglass baybuilt constructed by Ronnie Carman in Crisfield, Maryland, eleven years ago. The engine has over 14,000 hours on her. She is every inch a crab boat with room for twenty-eight pots on the canopy that runs the length of the boat and a "back porch" where twenty-eight more pots can be stored hanging out over the water behind the transom. Bushel baskets, some stacked to store the catch and others containing razor clams for bait, are on the deck along with more pots, crab pot buoys, or floats, and yards of line. The length of line he uses to attach a buoy to a pot depends on the depth of the water he is working.

Steve Smith prepares to put overboard a crab pot baited with soft clams. He won't change the bait in the pot if the clams are still alive, in order to save money, particularly when crabs are sparse.

We pull up alongside the first buoy. Steve snatches it up with a boat hook below the buoy and pulls it in by hand. In a month when the crabs start running hard and he's in deeper water, he will have another man on board to cull the crabs as he hooks the next buoy and wraps it around a winder, a round winch on the washboard beside the throttle, to bring it up quickly. Everything is synchronized, men and machines.

Right now, there are three crabs in the pot that he empties into a bushel basket. He checks the bait cylinder, grabs a handful of clams, crunches them in his hand, and stuffs them in. The pot then goes back overboard. "If you imagine a line from that marker off the shore behind us and run it all the way up to Holland Point, that's the line DNR set for us to crab outside of. If we place pots on the inside of that line they are illegal, and the crabs are always on the side that's illegal!" he states.

"Now, if you look at the line you can see that it's set so it doesn't run into Parker's Creek back there. I'm sure the crabs are or will move in and out of that creek so I want to set my pots in a line aimed right at it. I want 'em so they have to come by me to eat!"

I ask Steve about the placement of his pots. "Well, I get out here first in the spring, if I can, to stake my territory right alongside the line but the fact is nowadays there are no rules. A guy from the Beach [Chesapeake Beach] will come up here and put his pots right on top of mine. So I bought this boat so I can go in shallower, three feet, water and can move easier between the pots. Sometimes I use clams and sometimes fish for bait, depends on what they're interested in eating, but for the last ten years boys have been using clams so the [bait] buyers aren't purchasing any fish to speak of right now. Clams are cheaper and plentiful right now and at $25 a bushel they are less expensive than last year."

He stops to show me a female that has already shed. "This could be a good year. This warm spell helps and she's already shed. Looks promising. I can't tell yet what line the crabs are moving in toward that creek. If they start running in a pattern then I'll shift my pots a little accordingly. They want to get warm, and the water's still a little cold, so it seems they'll go in shallow, but damn it there's no crabs there. We're in seven feet of water and no crabs. It's still early yet. In the spring like this I'm the hunter–gatherer tryin' to get an edge, figure out where they are. In the summer we're all just gatherers haulin' them in like there was no tomorrow, if we're lucky."

He changes his tone when he tells me, "All the guys in my group have gone out of business just about. Now guys from the Beach and the Eastern Shore are crabbing here. They're good crabbers and can find the hot spots and that's why I like to lay my pots early as I can so I can stake my claim to a line. I hate getting started though. I've been off for five months and I've got to dip and paint pots, fix up the boat, and so on, but once I get started I make it a sport. Where are the crabs?!"

We've pulled up five large hardheads (fish) worth about fifty cents a pound but he doesn't keep them. "In the summer when we're working a lot of crabs, my crew man George, who's not a real waterman but he likes workin' with me in the summer, can really move. We're pullin', emptying, sorting, baiting, and replacing a crab pot every twenty-six seconds. And in this process, George will see where the crabs are in the pot. If he's got one or two at the top then they just came in and that's a good sign that they just swam in. For some reason, the last couple of years this spot off Deale has been a good, hot lick."

The sun, a red ball of fire crossing the horizon, sends a soft stream of light across the water and the boat. It is peaceful sitting on the engine cover talking with Steve as he continues to pull pots stretched in a line down the Bay. "Kenny Watts and other guys move around to catch their crabs. They're not as territorial as I am. A couple of years ago I would catch three bushels of females and Kenny or Billy Sherbo would catch twenty off Poplar Island. Don't ask me why."

There is no limit to the number of crabs a boat can harvest but the watermen are restricted to males that measure five inches from tip to tip of the shell and any mature female. In the spring and early summer the watermen are getting $90 to $100 a bushel. It is all based on supply and demand principles. In the fall when crabs are plentiful and people aren't eating them as often, they may only get $20 a bushel. For several years the Maryland Department of Agriculture's Seafood Marketing Program, under the direction of Bill Seiling, has spent a great deal of money to try to get people to eat crabs after Labor Day when they are fat, plentiful, and cheap. It has been a hard habit to break, and for some reason, for many regional residents, Labor Day continues to mark the last crab feast of the year. It's too bad because quite often the best crabs are up the Bay in September and even October.

Seafood marketing is big business in Maryland and Virginia, and the watermen like it that way. Steve and other crabbers will spend

$30,000 a year to crab, so making sure they have a market to sell to is critical.

Steve will use four bushels of bait. The clams should be only half alive, he says, or they will swim out of the bait cylinder. He doesn't use a door flap on his bait cylinder. "Some guys use doors all the time, even if they're puttin' in dead fish for bait. Not me, that's an extra step that I can eliminate. Drop 'em over right and the water will hold the bait in."

Steve's enthusiasm builds as the conversation shifts to his true love, the stock market, where he has been quite successful for over twenty-five years. He spends the next twenty minutes talking as if he's on the floor of the New York Stock Exchange, detailing company buyouts and histories, financing strategies, stock selections, pros and cons of high-technology stocks, bank stocks, and more. My head is reeling as if I'm stranded in a graduate school economics class and the professor continues to lecture over my head. Nevertheless, I am totally fascinated as I watch him pull pots at a slower pace because of the scant catch and listen as he moves from one financial topic to another. "I love being on the water more than just about anything, but I'll tell you if I'm not out here I'd rather be in front of my television watching the business channel and the stock reports all day long," he says. "We were all in Hawaii a couple years ago for vacation and Barbara gave me hell because I could barely pull myself away to go out on the beach!"

As we're talking—he's talking and I'm listening—we come across a buoy painted orange that's over the legal line. "I don't know him but see how he's not on top of me but, because he isn't, he's over the line and that infuriates me. First, because if a crab swims by he'll get it, and second, it's a $500 fine to be across the line and there's no cops around."

We are pulling up hard crabs now—not many, but a few—and as he empties them in the bushel basket, he explains that the peeler crabs (those going to shed) will make their big run around the first week of June. Steve will take the peelers and put them in water-filled shedding tanks on shore and sell them for soft crabs when they lose their shell. Crabs will shed all summer but tend to do so in cycles. They begin at the end of winter with the males shedding before the females. The second fin turns a white to pink color, indicating that the crab is about to peel, shed, and go soft.

"There's a lot of money in soft crabs, but it's a lot of work minding the tanks and getting the crabs as soon as they shed. If you have a lot of crabs in a lot of tanks it's a twenty-four-hour, seven-day-a-week job when they're in a cycle. It's brutal but we do it because we've got to make money when we can."

Most of Steve's crabs go to two crab houses that have been customers of his for many years. "I've had one for fifteen years, but I still have to fight for the price I want. Koreans run many of 'em now and some of the men don't deal too well with 'em. Me, I get along fine as long as they give me my price and keep buying from me.

"From now until the end of the season the first half of my day is catchin' the crabs and the second half is workin' the markets—seven days a week if I have to. I've got to keep them moving. When the crabs start shedding I'll leave the dock at 5:00 A.M.; come in around noon; change clothes; drive to town to sell the hard crabs; drive back to the yard to check my tanks for peelers and the soft crabs at 2:00 P.M., 6:00 P.M., and 11:00 P.M.; go home in between and after the last check; sleep until 2:00 A.M. then get up to check them again; go back home and sleep; and get up at 4:30 A.M. to go back out on the water again. Ethnic groups tend to like females, and they're used a lot in soup. Crab houses and so forth want the males. My job is to keep 'em in crabs no matter what they want."

It is an understatement to say that Steve Smith is a competitive man. Everything he says relates to time, efficiency, business policies, and money. "I love what I do. I had to admit to myself recently that I'm not as passionate as I was when I was young because now I work just for the money. I try to do more today than yesterday—a successful business growth trend, right? If that doesn't happen then I have to change the way I'm crabbing, and change it for the better. I put a lot, perhaps too much, pressure on myself and that's a problem I have to deal with," he says.

Steve lays his pots north to south and checks the line to see how the crabs are sitting in them. He thinks they are lining up with Parker's

Creek and, if he's right, he says he will rearrange some of his pots accordingly. He and other crabbers around the Bay will do this frequently as the crabs move and they are looking for hot licks, or areas where crabs are plentiful. They stay in an area until they see a pattern of movement. "This has been a good spot for several years now if I work in fourteen feet of water," he says. "An Eastern Shore man picked up 150 pots and moved yesterday. I'll stay a few more days because I'm doing pretty well and it's early yet."

In spring the crabs generally live in shallow water from three to fourteen feet deep, where the water is warmer. In the fall when the whole of the Bay and its tributaries are warm, the crabs are found in water thirty to fifty feet deep. Steve will analyze the catch in October. "Last year at that time of the fall I was working water eighteen feet deep and I'm pretty sure they went in the mud there. Then in the spring they come out and move to shallow, warmer water and I'll catch 'em there. I may be wrong but that's the way I figure it," he says.

Watermen keep written and mental records: catch, money, date, and location. They look where the crabs may stop to eat, an edge, a channel aimed at a creek, and so on. "After twenty-seven years at this I'm still fine-tuning the effort. They are tough to figure out and I use every resource I can to try to win this game," he says.

Watermen are allowed by regulation to have nine hundred pots with an escape ring in the wire to let smaller crabs out. They must take a day off a week, and there is limited entry, meaning that there is a set number of crab licenses issued each year. Men and women are "grandfathered in" and issued another license if they have had one before and intend to continue, so there are few new licenses available.

Regulations exist to try to pacify those who think crabs are in a decline, but Steve is not sure if these regulations help the crabs at all. "We can have good years or bad years with the regulations. We can have good years and bad years with plenty of rockfish that eat crabs. We know rockfish and weather impact on the crabs but I don't think anyone really knows anything more than anyone else does. Overdevelopment and growth have impacted the water quality of the Bay and clearly have had a negative impact on the natural environment. But if you ask me about that impact on the crab population, except for the fact that we've lost too many of the underwater grasses, which is bad, I couldn't tell you. Last year some of the boys had a record year and it was the worst year I've had in all my time out here. And I'll tell you crabs don't like dead bait; they like it fresh. But a couple of years ago when bait was real expensive I left it in for a couple of days, and guess what? I caught crabs. So, you tell me what the answer is. I'll tell you one thing for sure: You can't figure them out and you can't outsmart them. You just get lucky, I believe.

"Sometimes crabbing is like stealing money, but if it was easy everyone would do it and boys wouldn't leave it. It's hard work and I've lost a step as I've gotten older. You can't mind being alone out here for hours. I love it because I don't like to socialize that much anyway. I don't socialize in the waterman community except for my group, and I don't get involved in the association. Maybe I should but I'd rather talk stocks all day, and the way I look at it we're all after the same crab. This is open range out here. I can't for the life of me figure out why someone would tell someone else, 'Hey, they're biting over here. Come on over and take some of my catch.' "

As we work the water off Broadwater Creek he notes that we are in a shallow spot and with the wind shifts it will get rough. "We're gonna move inside where it's calmer. No sense in beating ourselves up this early in the season."

The pots he will use to attract peeler crabs have smaller mesh to catch smaller crabs. The crabs will shed twenty times in a lifetime in this part of the Bay—more frequently down south where the water stays warm longer—and a life cycle is about two years. Steve paints all of his pots and attaches zinc bars to the rods to keep algae and rust from accumulating. "I do hate doing that and I try to put it off as long as I can. The new owner of the yard didn't want us painting there anymore so I do it at the house now. Lots of these pots are Scottie's and need work. I'll try to make them useful as long as I can."

The crabbers aren't seeing real runs yet and may not until mid-June after they begin to shed. Then a man can make $2,500 to $5,000 a week selling soft crabs with his hard crab catch. The watermen will generally make good money in June and July. The crabs will drop off in August, if they live up to recent historical actions, and pick back up a little after Labor Day.

As we head home with two bushels of crabs, Steve feels pretty positive, despite the fact that we didn't catch a third bushel or see a real pattern to the crab run. "The boys that got out here early in April got the first run of crabs from the mud. We're just feeling our way now. It's a battle of wits. I've seen some big crabs today. I've seen some peelers. I've seen a little movement. It looks like a good season is coming soon. Check with me in a month and see how we're doin'."

Back at the dock after the crabs are put in a refrigerator and everything going ashore is put away, a mangy old red cat comes around the corner of the box. "That damn cat was nearly dead years ago. Look at his side. It's half eaten away with cancer or something. He comes around every day when I come in and I give him a fish and talk to him a little. I can't believe he's still alive. But there he is," Steve says with compassion in his voice, and I am reminded of another seemingly tough, independent, slightly arrogant waterman named Danny Beck in Baltimore, who has his blue heron to greet every day and, of course, to feed him a fish. Under the tough waterworn exterior is a tender heart—maybe not for regulations, marine police, the trappings of urban living, or yuppies at a cocktail party—but certainly for a puppy, a crippled cat, or an old blue heron, and that's good enough for me.

Back at my house, Steve wants to see our renovated kitchen. He comes in and we chat about home improvements for a few minutes. To me he is usually a little reclusive and I wonder if somehow I broke through to him. We'll see, I think, as he finally walks to his truck, shakes hands, and invites me to go out again anytime. I will take him up on the invitation.

CRABBING THROUGHOUT THE SUMMER

Crabs remained relatively expensive through the first weeks of June, as the crabs moved up the Bay and crabbers began adding peelers and then

Half-filled crab baskets can be an early sign of a rough season ahead. Crabs, as unpredictable creatures, give watermen fits of frustration because their populations, locations, and weight can change quickly. One old-timer said, "I can chase a crab all day and never figure out where that son of a gun will be next." Courtesy Maryland Watermen's Association.

soft-shells to their lucrative market. When I bought crabs from family friend Jeff Hickman of Bayside Crabs in Deale, I paid $120 a bushel for large males, or jimmies. His peeler/soft crab operation was getting in full swing with the soft crabs selling for about $2 each. This was the time of the season that Steve Smith talked about figuring out which way the crabs would travel and making good profits off increasing catches.

Unfortunately, June 13 is cold and overcast with drizzle falling on Russell Dize's dock on Tilghman Island where he waits for the last of the crabbers to pull in and offload the day's catch. The middle of June, which is traditionally a prime crabbing time, is starting to look as dismal as the weather.

"I think the fish are eating the crabs. That's got to be it," Dize surmises. "That's got to be it," he says as he leans back in his chair at the small desk where he keeps notebooks of catches going back many years. "We were getting terrific catches earlier, but once they shed, we're getting nothing. I can't nearly fill my orders from vendors."

As *Julia Rose* pulls in with Captain Jack Thompson at the helm, the men who wait to unload move toward the boat. Thompson has one and a half bushels to hand over. He shakes his head. "If crabs were roosters, they sure aren't crowing nowadays," he says as the men laugh. Crabbers are earning $150 a day for their catch. Normally, in this high time of the season, a crabber can earn up to $500 a day with a nice haul. The atmosphere on the dock remains one of good humor and expectation. The numbers have to get better. They can't get worse.

Then, as everyone in the region prepared for Fourth of July festivities that more often than not included crabs, the fishery dropped off even more and crabbers were having a difficult time meeting demand. Kenny Keen, Charlie Quade, and others pulled their pots for the rest of the season. It was too expensive to fish for no money. To make matters worse, on June 21, the Chesapeake Bay Commission's Bi-State Blue Crab Advisory Committee, comprising lawmakers from Maryland and Virginia, met in Annapolis to hammer out a plan to better manage the suffering crab population, including a discussion on limiting the harvest in both states. Crabbers said they were making 25 percent less money crabbing than five years ago, making it nearly impossible for new generations to afford to work in the industry. Others said that limiting the harvest would put small operations out of business.

Maryland Delegate John Wood, Jr., chairman of the committee, said, "We have many challenges facing the crabbing industry." He emphasized that the two state governments in 1999 each put $150,000 toward a two-year analysis of methods to better manage the crab's $32 million industry.

Then, on June 27, the Virginia Marine Resources Commission voted to declare a vast 660-square-mile area of the Chesapeake off-limits to commercial crabbers during the summer months, in hopes of saving ten to twenty million female crabs a year and reviving what some consider to be a faltering fishery.

Beginning July 1, the sanctuary stretched like a giant crab claw from the mouth of the Chesapeake to the Maryland border, roughly one-third of the fishable waters of the Bay and its tributaries in Virginia. The commission, in an unprecedented move of this magnitude, ruled that for three years this area would be closed to commercial crabbing from June 1 to September 15, the time when female blue crabs swim to the warmer, saltier mouth of the Bay to spawn.

"We may be teetering on the edge," Rom Lipcius, of the Virginia Institute of Marine Sciences, told the *Washington Post.* "You never know when a fishery is about to collapse until after it's happened."

Seasonal sanctuaries are an increasingly popular worldwide technique to protect fisheries. The Georges Bank off New England has imposed a sanctuary on sea scallops, and the Florida Keys have a small sanctuary for all marine life. Over the past decade Virginia's commission tried to impact the harvest of crabs by issuing a freeze on crabbing licenses and by taking other measures, including the creation of a small sanctuary near the mouth of the Bay in 1994.

For years, environmental groups, and particularly the Chesapeake Bay Foundation, have applied relentless pressure on Maryland and Virginia regulators to clamp down on commercial crabbing despite opposition from the crabbing industry. While most crabbing takes place in shallow waters, some Virginia crabbers move to deeper water in an effort to intercept the migratory movements of the females (the pregnant females, called sponge crabs, are protected) that can be caught in far greater numbers in their region of the Bay. "Creating the sanctuary is probably the best thing we can do to head off increasing pressure on crabs in that part of the Bay," said Robert D. Brumbaugh, a scientist for the Chesapeake Bay Foundation. Scientists further stated that to impose a similar sanctuary across the border in Maryland would be impractical because the deeper reaches of the upper Bay do not have enough oxygen to support crabs. Further, Maryland crabbers tend to take their share of the two hundred million crabs harvested each year in males which are partial to the cooler, less salty water of the upper Bay. Virginia crabbers take mostly the abundant females, particularly toward the saltier, warmer mouth of the Chesapeake.

Reaction from crabbers regarding the sanctuary ruling was mixed. Some contend that it is unnecessary, blaming the declines of the crab harvests on rebounding numbers of rockfish and other predators that feed off young crabs. Recent droughts and hurricanes may also have shifted water salinity in ways that have hurt crab harvests, they said. Richard Stilwagen, of the Virginia Watermen's Association, said, "We believe that once the sanctuary is established, it's going to open the door to a continued expansion of that sanctuary. It's the commercial fisherman that's always taking the hit."

But some watermen, sensing that their way of life may be in danger, took a different approach. "It has a lot of potential to increase the crab population statewide which will benefit all watermen," said Doug Jenkins, Sr., of the Twin River Watermen's Association in the Northern Neck region of Virginia.

Regardless of anyone's position on the decline of the fishery or the effectiveness of the sanctuary concept, crabbers are impacted either way. For many of them, these are red flags signaling that crabs may become an endangered species, despite the fact that throughout the nineties catches averaged about 42 million pounds per year.

While criticisms abound about commercial crabbing, it interests me that in 1993 the number of noncommercial crabbing licenses issued was more than double the commercial licenses. Now no one knows just how many crabs these "chicken neckers" catch, but it has to be a significant number; in 1988 the recreational harvest was estimated at 21.5 million pounds. Nevertheless, in 1994 harvest pressure remained, and for the first time since 1972, restrictions were placed on the number of crab pots allowed per professional crabber. The limit was three hundred pots per license, although additional allocations can be purchased for one or two crew members, not to exceed nine hundred pots per boat. Cull rings, to allow small crabs to escape, were required to be installed in all pots. Limited entry gave the state the authority to establish a prescribed number of people to participate in any given fishery; reporting catches became mandatory for all commercial harvesters; restricted hours were imposed on commercial gear types, including lengths of trotlines; and new restrictions were added every year until the millennium. Watermen spend a great deal of time talking about and analyzing the numerous restrictions. They must, because after reaching maturity crabs live an average of one year, and rarely more than two. That's not a whole lot of time for the hunter or his prey.

On August 23, 2000, the headline in the Annapolis *Capital* read "Saving the Blue Crab." The article followed over four weeks of extremely poor crab harvests and watermen moving from this typically successful catch to other fisheries to try to make a living during a time when they are normally earning a lot of money.

Experts met with the Maryland House Environmental Matters Committee following a summer study of Maryland's famed blue crab industry. In reaction to the shortage of crabs, smaller crabs, and prices reaching elevations few have ever seen—$195 a bushel at Annapolis Seafood Market—Delegate George Owings, III, a neighbor and friend of the watermen, said, "We're searching for answers here. They're hard found, I have to tell you."

Even though prices would drop after Labor Day as crabs become more plentiful up the Bay, people were wringing their hands with anxiety, and it was not just the watermen. Maryland leaders recognized that at least two-thirds of the state watermen's income is derived from the blue crab catch. Doug Lipton, with the University of Maryland Department of Agriculture and Resource Economics, formerly Sea Grant, is someone on whom I have relied for over a decade when I want to know the reality of an issue dealing with recreational or commercial economics. He told me plainly, "As we move into the millennium and look at trends this summer, if the blue crab were to go away, the fishing industry in Maryland as we know it would go away."

Watermen have known for years that the blue crab is difficult "to figure out, at best," and scientists confirm the crab's elusiveness. Jonathon Zohar, director of the University of Maryland Biotechnology Institute, observed, "It is really surprising how little we know about the basic biology of an organism of such economic importance."

As tensions mounted, watermen complained about recreational crabbers' taking more crabs than commercial fishermen do, about too many fish to support the crab population, and about pollution. One thing is certain: only 3.2 million pounds of crabs were harvested in July, less than half the average for that month. Further, the harvest for the first four months of the eight-month season hit a record low of 9.7 million pounds, approximately one-third lower than the average. That means that watermen were earning 25 percent less money crabbing than they were five years ago. Balance that with inflation, cost of equipment, gas, bait, and time and it is an abysmal situation.

Virginia instituted its sanctuary covering much of the southern Bay and making it off-limits to crabbing during the summer of 2000. Maryland would not follow that lead and decided to wait until the spring 2001 completion of a two-year, $150,000 study conducted by the Bi-State Blue Crab Advisory Committee. Bill Goldsborough of the Chesapeake Bay Foundation, who has complained about overharvesting for over a decade, said, "Overharvesting and loss of Bay grasses have contributed to the decline. As crab populations have suffered, crabbers have stepped up their efforts to catch them and we're stretching the limits of the crab population."

Bob Evans, whom I accompanied on hand tonging and catfishing trips during the winter, also sells crabs retail and wholesale, using his own boat and buying from other watermen. Evans believes recreational crabbers are having a far greater impact on the harvest than scientists will admit. The plight of the crab harvest this summer has forced him to move his boat from Shady Side in his native Anne Arundel County up to the northern reaches of the Bay, to the Susquehanna Flats, where crabs are a bit more plentiful.

While recreational crabbers have regulations on the number of pots they can put out, these laws are relatively impossible to enforce, especially since they don't have to report their catch. State officials estimate that the chicken neckers could be pulling as many crabs from the Bay as the commercial watermen.

"I think everybody has a God-given right to go out and catch crabs to eat," Evans says, "but recreational crabbers have the ability to catch two bushels a day. They don't need two bushels a day and we do. There needs to be a level playing field here and we want tighter restrictions placed on the recreational people."

"So many things you could say have caused the decline, but we really just don't know," said J. R. Gross, a fifth-generation waterman from Churchton, Maryland. "I just know it's bad. Really bad." Gross, like other watermen, turned from crabbing to fishing other species when he couldn't make a decent wage.

With successful stocks of hungry large croaker and rockfish, cooler temperatures, and too much rain, it appeared that the 2000 crab harvest was far below anticipated goals of watermen and scientists alike. The 2001 season is expected to be equally or perhaps even more unsuccessful. With this anticipated disappointment, Phillips Seafood, the MWA, and the state have provided funding for research at the University of Maryland Biotechnology Institute on crab hatchery techniques, with a particular focus on attempts to spawn female crabs out of season in a laboratory situation. After three months, approximately four thousand baby crabs are one to one-and-a-half inches long. "This could be a cutting-edge breakthrough for us," said MWA's Betty Duty.

"With this scientific progress, the state's crabbing regulations still have not been finalized as of April 2001, but regulations should be done and implemented by May 1 because the season started April 1 with men working under 2000 rules," Betty added. "Virginia is waiting to see what Maryland does. All in all, they seem fair to everyone."

CHAPTER 12

Hope for the Bay

Children are sometimes guilty of picking on the smallest kid in the neighborhood or targeting an easy mark for the brunt of childhood pranks. In the animal kingdom as well, the weakest is usually the easiest prey. It seems to me that the watermen and their subculture, or commercial fishermen in general, are more often than not the victims of undeserved criticism, regulation, or legislation by society at large because the fishermen seem to be easy prey and want to avoid difficult situations.

These men and women watch with cautious optimism and, perhaps, skepticism, as millions of dollars are invested to save the Bay. They watch the reports on declining species, overharvesting, shifting bureaucratic approaches, political posturing, new management attempts to revitalize fisheries, and more. When reports contain recommendations that are followed, the consequences are often at the watermen's expense. Biased news reports, often substantiated with manipulated statistics, frequently point fingers at the watermen for overharvesting or for sacrificing the health of a fishery for their own immediate gain. All too often, however, those making changes fail to recognize that, as we invest millions of dollars seeking a better way to protect the watershed, the watermen—our native hunter-gatherers—are consulted about such legislative issues only in a cursory fashion. Usually, instead of being able to contribute to a solution, they are forced to defend their rights or their way of existence. This is unfortunate because we frequently fail to capitalize on what the little guy, the easy mark, can show us.

Watermen know best when species are declining, improving, or thriving. They know when eels move from the mud; that crabs move to different sites following water temperature, salinity, and currents; when and where oysters are surviving or dying; when and where rockfish are maturing beyond the juvenile stage; where grasses are growing or receding; and so much more. They don't need expensive reports to tell them these things. They know them because their existence depends on that knowledge. In addition, they can anticipate change by looking at the shoreline and seeing new homes being constructed within the one-thousand-foot buffer established by the Critical Areas Commission over a decade ago, because politics and money answered the developers' call. They also watch a recreational marina expand its amenities and increase its slip rates, which will exclude them; or a new sewage treatment plant that essentially kills everything in a wide radius of its flow; or run-off from chicken, cattle, or hog farmers who are not following best management practices to protect small tributaries that flow into a river and into the Bay. Unfortunately, as I see it, the waterman is "the little guy" who often can't fight back, so he tries somehow to survive the consequences of other people's actions.

DUMPING IN SITE 104

To say that all hell broke loose would have been an understatement when, in 1999, the state really started pushing a plan to dump dredge spoil from the Baltimore Harbor shipping channel into the Chesapeake off Kent Island at Site 104. Environmentalists, watermen, residents—particularly those from Kent Island, led by former waterman and now county commissioner George O'Donnell—and politicians spent the rest of that year and the beginning of the 2000 legislative session either fighting or supporting the proposal as factions broke off from all segments to take sides and express their opinions.

Maryland House Minority Leader Robert Kittleman, a Republican from Howard County, revealed his opposition to the plan when he said, "We have laws to prevent sediment runoff, yet the governor wants to intentionally dump eighteen million cubic yards of sediment into our Bay." But Governor Parris Glendening's administration still pushed its proposal to dump the Baltimore spoil into a four-mile deep trench called Site 104, insisting the site was a vital part of the state's twenty-year dredging plan to keep the Port of Baltimore competitive with deeper water and a wider channel. The port and related firms also employ thousands of Marylanders. The governor could not ignore those people or their votes, and that is a difficult decision for any politician to make, regardless of his or her true feelings about the project.

Bill after bill was introduced in the legislature to ban open Bay dumping, and even members of Maryland's congressional delegation were caught in the fray. Congressman Wayne Gilchrist, a longtime friend of the watermen, opposed the dumping and helped force the Army Corps of Engineers to reevaluate the impact on the Bay. Then a Mason-Dixon poll showed that 67 percent of state residents oppose open-water dumping of dredge spoil, whether they understand the issue or not.

State Senator Robert Neall, a Democrat from Davidsonville, revealed that he had, perhaps, the most realistic response to the entire issue, when he said, "I think we learned from pfiesteria that we can't make good decisions without good science, and I've seen a lot of opinion about Site 104 but not a lot of science."

In May, MWA President Larry Simns decided it was time to clear up confusion surrounding a variety of issues involved in Site 104 and its relationship to state funds for oyster seed planting programs. He emphasized that Site 104 had been used as a dump site for Baltimore's contaminated spoils in the past. The MWA, despite opposition from some of its members, felt that the new dredging proposal should consider Site 104 for several reasons, Simns explains: "It was already contaminated; the bottom was already damaged beyond hope; and there is still room in it. Most importantly, it was promised that the contaminated material would be covered up, the bottom would be smoothed out, and it would be made workable once again."

For Simns and the MWA leadership, the tough decision was based on the fact that watermen would receive a dollar per cubic yard of waste that would be dumped. This would save a lot of money because the port authority would dump at a site closer to the harbor than the intended sites such as Poplar Island and Bodkin Creek. The extra money from the port authority budget for keeping channels open went into the oyster replacement fund to prepare to build life in other areas to compensate for the destruction—mostly of worms—on the bottom of Site 104. "To us, the rehabilitation of an oyster bar is worth much more than what had been destroyed on 104, and as the destruction was supposed to be temporary, we decided that the trade-off was reasonable," Simns explained.

To some this seems ironic, as Simns and the association have been the major force fighting to put an end to open-water dumping, particularly in other areas like Pooles Island. Simns has a reputation for looking at the big picture, and this time the money received from dumping in an already bad site would save the oyster seed planting program. "I didn't hear one suggestion for an alternative site. If there was one waterman who could illustrate that the trade-off was not worth the end result, then I told them to talk to me. He needs to show me that the damage will outweigh the benefit. It takes a million dollars per year for Langenfelter to set up the [oyster seed dredging] operation and haul shells to the seed bars. So, right off the top comes a million dollars that goes to digging, hauling, and putting shells on the seed bars. None of this would be possible without the money from Site 104.

"Additionally, some of the money," he continued, "other than the $1 to $1.5 million that goes to the shells, is designated for experimentation with hatcheries and the hatchery oysters. We can't sit back and ignore the new technology. We've got to make hard decisions and know that we're not making a sacrifice for other groups; we're making a sacrifice for ourselves. We must keep the seed program alive and if it takes money from dumping into Site 104 to guarantee it, then that's the hard decision we have to make."

By the end of August 2000, in spite of support from Simns and the MWA, Site 104 was no longer being considered as a site for the dredging spoils of 18 million cubic yards of Baltimore Harbor, or for the proposed dredging of the Chesapeake and Delaware Canal and maintenance and reshaping of the Tolchester "S" turn, a winding portion of the canal route to the Port of Baltimore. With all levels of government involved in the state's dredging issue and then with the U.S. Army Corps of Engineers finding that dredge material contained contaminants toxic to several fish species, the governor decided to withdraw the plan to dump at Site 104.

The Maryland Port Administration has recently proposed the dumping of 80 million cubic yards of dredged mud and silt into any one of ten possible sites, which would create a new island in the Bay. The island would range in size from seven hundred to three thousand acres and cost up to $800 million. Proposals of this type have come to fruition before, and islands like Hart-Miller, Pooles, and Poplar were created. Some were completed above the water line while others were turned into oyster bars. As with any proposal affecting the Bay, environmentalists and locals living near the proposed sites have launched strong opposition efforts.

Former oysterman and president of the Queen Anne's County Commission George O'Donnell was one of the first to oppose Site 104. He is concerned about the island plan but "obviously, if we could do that without putting an island near our county that would be our preference. But we may have to deal with that for the salvation of the port."

With the issue of dumping in Site 104 forgotten, new proposals will emerge, some focusing on the solution of creating more islands. Marylanders will fight their battles, taking sides to preserve the economic benefits of the Port of Baltimore or to foster the ecological improvement of the Chesapeake.

STATE OF THE BAY AND THE CHESAPEAKE 2000 AGREEMENT

Thirteen years ago a young, enthusiastic woman working grassroots development for the Chesapeake Bay Foundation took on a new job as executive director of the Chesapeake Bay Commission. The tri-state legislative commission was created by the federal government in 1980 to advise the members of the General Assemblies of Maryland, Virginia, and Pennsylvania on matters of Baywide concern. Little did my friend Ann Pesiri Swanson, now forty-three, know how dramatically her professional life would change. Swanson would be the staff person responsible for leading the direction of the commission on issues addressed by its members that would be as wide-ranging and complex as the Bay itself, delving into matters of air, land, water, living resources, and the integrated management of all of them.

Over the years Swanson and her staff would be tested in the most unique nuances of human, government, and public relations as they cajoled, lobbied, pushed, and pulled twenty-one members of the commission from three states—including fifteen legislators, three governors' cabinet members, and three citizen activists—to finally pull together Chesapeake 2000: A Watershed Partnership, otherwise known as the final Chesapeake Bay Agreement. Swanson is given credit by environmentalists, government officials, and watermen for almost singlehandedly drafting the agreement as she led the program on an eighteen-month journey, word for word, through the draft document, making sure that it didn't crumble beneath its own political weight throughout the process. On June 28, 2000, Chesapeake 2000 was signed by the Chesapeake Executive Council, three governors, the mayor of the District of Columbia, the chairman of the Chesapeake Bay Commission, and the administrator of the U.S. Environmental Protection Agency, to serve as a guide to Bay restoration efforts for at least the next decade.

According to Swanson "what is very unique about the final agreement is the fact that it was truly a bipartisan effort. In the past, these efforts were driven by Democrats—but not any more. And this agreement is very specific with nearly a hundred commitments to action over the next ten years as we go beyond the Bay proper and move upstream right into people's backyards." The dedicated members of the commission, the leaders and staff of the Chesapeake Bay Program, and the hundreds of private citizens, bureaucrats, scientists, watermen, business

On June 28, 2000, the Chesapeake 2000 Agreement was signed. *Left to right:* Bradley M. Campbell, Regional Administrator, U.S. Environmental Protection Agency, Region 3; James M. Seif, Secretary, Pennsylvania Department of Environmental Protection; Anthony Williams, Mayor, District of Columbia; Parris Glendening, Governor of Maryland; John Paul Woodley, Jr., Secretary, Virginia Department of Natural Resources; Bill Bolling, Virginia State Senator.

owners, and others who participated in meeting after meeting over many years deserve tremendous credit and the region's deep appreciation. Reaching the final signing of Chesapeake 2000 was a monumental task. In this book, it would be impossible to address all of the issues involved in the agreement and the actions to be taken, but readers are encouraged to contact the Chesapeake Bay Commission office in Annapolis, Maryland, to ask for a copy of its Annual Report 2000: Keeping the Agreement—an astounding piece of work. Copies of the agreement and a summary of its findings are also available.

The agreement sets new measurable limits on sprawl and carries them across state borders for the first time in the history of the nation. The new Chesapeake 2000 agreement also calls for stepped-up efforts to control nutrient and sediment pollution, to set new catch targets for blue crabs by 2001 (the commission created the Bi-State Blue Crab Advisory Committee in 1996) to bring the fishery back from its threatened state, and to increase educational efforts to help residents throughout the watershed understand their role as Bay stewards. We must all be caretakers, responsible for helping to improve the water quality of the Bay. In a very progressive and positive move, the leaders made a commitment that every student will have some experience with the Bay before graduation, through field trips and classroom study.

The goal to reduce sprawl is critical, with an estimated three million more people expected to live in the watershed over the next twenty years. The conversion rate of resource land most likely will be measured using data from the U.S. Department of Agriculture's Natural Resources Inventory, which is conducted every five years. Preliminary estimates indicate that about 110,000 acres of land are developed annually in the three states' part of the watershed. If that figure holds true, the new agreement would, in effect, limit development to 77,000 acres a year beginning in 2012.

Aside from curbing development, the agreement seeks to preserve open space by guaranteeing that one-fifth of the watershed will be permanently protected through easements, land trusts, or public ownership. That would mean increasing the protected area from the current preservation of 17.2 percent by 2.8 percent to meet the 2010 goal. This will require the permanent preservation of an additional 1.1 million acres, an achievable goal considering that 6.6 million have already been preserved.

The watermen are most interested in those portions of the agreement that pertain to the main fisheries of the Bay: the tenfold increase in oysters, maintenance of the menhaden population to conserve striped bass, and the effect of striped bass predation on the blue crab population. The agreement calls for the development of multispecies fisheries plans by 2005 that will take these issues into account by targeting various fisheries instead of only one species, like the rockfish moratorium of 1985.

Further, watermen and environmentalists have expressed their concerns, often vehemently, over the continued disappearance of submerged aquatic vegetation over the period from the sixties to the nineties. The underwater grasses are one of the most important habitats in the Bay for juvenile fish, waterfowl, blue crabs, and other species. Increased algae blooms and sediment cut off sunlight for the plants, causing a devastating decline. To bolster this effort, the agreement calls for restoring 25,000 acres of wetlands by 2010 and setting a new goal for planting streamside forest buffers by 2003. That will go beyond the present goal of planting 2,010 miles of forest buffers by 2010.

The agreement describes nearly one hundred actions aimed at achieving a Bay filled with "abundant, diverse populations of living resources, fed by healthy streams and rivers, sustaining strong local and regional economies and our unique quality of life."

The agreement was signed at Herrington Harbor Marina in Rose Haven, Maryland. The ceremony drew more than seven hundred people as witnesses. An obvious concern, however, is whether this agreement is valid or if it is being used as a shallow means for politicians to ensure reelection or for others to simply bill themselves as proponents of the environment. We can only wait to determine whether the commitment to make hard decisions despite political pressures will be a truly historic legacy of the signers. In the past, money from developers and others has influenced politicians, resulting in the disappearance of commitments such as those contained in the 2000 agreement. Mother Nature, however, has proven herself a very powerful force, providing us with hope for the future of the Bay. At the time of the signing, it rained on the tiny community of signers and witnesses—a good sign after two years of drought in the watershed.

As the Bay Commission sets its standards for commitments and states pledge their legislative support each year, the private Chesapeake Bay Foundation presents its own State of the Bay report. While some watermen consider this report biased toward the accomplishments of CBF and its members, it does reflect the foundation's appraisal of the condition of the Bay.

The 1999 report analyzed twelve categories and, on a scale of 0 to 100, the Bay's health rated a 28, one point better than in 1998. To determine this number, CBF compared the Bay as it is today against the healthiest Chesapeake we can describe—the rich and balanced Bay that

Captain John Smith portrayed in his exploration narratives of the early 1600s. While the Bay bottomed out in the health reports in 1983 with a rating of 23, the work of public agencies, private groups, and tens of thousands of volunteers has helped bring about a slight improvement. Consider the CBF analysis of the Bay, released in 1999:

Habitat

Wetlands: 42 points (−1 from 1998). Thousands of acres of wetlands have been destroyed in Virginia since 1998 as a result of a court ruling that reopened a loophole allowing the ditching and draining of wetlands.

Forested buffers: 53 points (no change from 1998). Of the basin's 110,000 miles of streams and shorelines, 53 percent have buffers.

Underwater grasses: 12 points (no change from 1998). The drought of the past two years may have slowed runoff and allowed for underwater grass growth, but gains in the upper Bay tributaries were offset by drastic decreases in Tangier Sound, which has lost 62 percent of its beds since 1992.

Resource lands: 33 points (no change from 1998). The Bay's watershed is continuing to be converted from open space to developed land at an alarming rate.

Fisheries

Rockfish: 75 points (+5 from 1998). Scientists must watch the menhaden fishery, a staple of the voracious rockfish, and the possible decrease in the population of older, larger fish that are needed to maximize spawning potential. Overfishing remains a concern.

Crabs: 48 points (−2 from 1998). Only the crab's natural resilience prevents a lower rating as continued pressure from heavy fishing and a decline of submerged grass beds impact the state of the fishery.

Shad: 3 points (+1 from 1998). The population remains depleted, but strong interest in restoration of the fishery has sparked some optimism.

Oysters: 2 points (+1 from 1998). The population continues to suffer from disease, irregular reproduction, and a lack of suitable habitat. Growing and transplanting over ten million oysters onto reefs throughout the Bay provides hope that the ratings will continue to improve.

Pollution

Toxins: 30 points (no change from 1998). CBF believes that both widespread adoption of "zero discharge" goals and commitments to actual reductions in the use of toxic compounds are needed to significantly reduce the levels of toxics.

Nitrogen and phosphorus: 16 points (both +1 from 1998). Continued input from air pollution, sewage treatment plants, and agriculture and the lack of a Baywide trend toward reduced levels give little hope that this rating trend will change in the coming years.

Water clarity: 16 points (+1 from 1998). Clarity is an indicator of nutrient and sediment loading to the Bay. The drought helped reduce runoff of sediment, but agriculture and development continued to add nutrients as we continue to see a decline in forested buffers, wetlands, and underwater grasses that filter nutrients before entering the Bay.

Dissolved oxygen: 15 points (no change from 1998). This rating is a result of the nutrients entering the Bay.

In addition to providing the ratings in the previous key catagories, the foundation also called on the Chesapeake Bay Program, the government partnership working to clean up the Bay, to set the following goals in each of the twelve areas:

Wetlands. Reduce the annual loss by 75 percent so that 125,000 acres may be restored by 2010.

Forest buffers. Reduce the annual loss by 75 percent so that 5,000 additional miles may be restored by 2010.

Underwater grasses. Cover 225,000 acres in the Bay and its tributaries by 2010.

Dissolved oxygen. Increase oxygen levels by at least 5 parts per million in spawning and nursery areas and at least 3 parts per million in other areas by 2010.

Resource lands. Reduce current land development by 50 percent to 45,000 acres per year and protect an additional 500,000 acres of farms and forests from development by 2010.

Toxins. Reduce 1995 levels by 50 percent by 2010.

Water clarity. Improve so that it supports underwater grasses at a depth of 6 feet by 2010.

Nitrogen and phosphorus. Permanently reduce 1985 levels by at least 50 percent by 2010.

Blue crabs. Restore the population to the size and composition of the 1960s by 2010.

Rockfish. Increase the older and larger fish population by 2010.

Oysters. Set aside 10 percent of the traditional oyster grounds as sanctuaries that incorporate 1,000 acres of rebuilt reefs. And, for the watermen, another 10,000 acres of oyster grounds in the vicinity of the sanctuaries should be rehabilitated as harvest bars by 2010.

Shad. The ten largest tributaries should each have spring shad runs sufficient to support fisheries without jeopardizing the continued growth of the shad populations.

Despite the goals of CBF's State of the Bay Report, many among the 15 million regional residents express concern for the viability of the Chesapeake Bay's natural resources. When the agreement was signed on June 28, 2000, government leaders in the Chesapeake area took a major step in favor of the Bay, and the watermen acted as witnesses. Perhaps Will Baker, president of the Chesapeake Bay Foundation, summed it up in the most realistic light: "The Bay will only get cleaner if the commitments in this agreement are actually kept. We call on the signatories to develop the programs needed to convert the words in the Bay agreement to improvements in the Bay ecosystem." Watermen, in the meantime, will wait on the sidelines to see what new regulations will limit their economic viability in the name of conserving resources.

A PLACE TO TIE UP

Near the end of June 2000 approximately 150 people gathered at Solomons, Maryland, to mark the creation of a dedicated county-owned wharf for commercial watermen in Calvert County.

Linda Kelley, president of the board of Calvert County Commissioners, said that for 350 years the core of the county's economy and its culture have rested on two pillars, agriculture and fishing. She had begun working with the watermen in 1998 to locate waterfront property where they could dock their boats with suitable facilities for loading and unloading them and for hauling their catch to market.

At the dedication, Kenny Keen, president of the Calvert County Watermen's Association, said, "We have needed a place like this for fifteen years. The population growth in Calvert [County] and the increasing popularity of recreational boating has placed increasing pressure on marina owners to keep commercial men out. Slip fees have gone up to the point that working watermen can no longer afford them. Recreational boaters didn't want us around anyway, with noise, smells, and long hours that we have to work. We have had to endure hurricanes, tropical storms, moratoriums, oil spills, too many regulations, and we're still here."

Chesapeake Beach waterman Bobby Abner added, "A dedicated wharf in Solomons is going to save me hours on the Bay when I am fishing the waters around the mouth of the Patuxent River and south." The wharf is located on property purchased five years ago by the University of Maryland's Chesapeake Biological Lab. The university will build a new lab at the back of the property and the watermen will use the waterfront. The county has a ten-year lease with an option to lease for another ten years. To complete the facility, the county and Maryland's Department of Natural Resources will be investing $200,000 in upgrading the docking areas.

On July 21, waterman Bob Evans called to ask me to join him the following day for a meeting with Adam Hewison, a computer entrepreneur who recently purchased the Chesapeake Bay Institute property in Shady Side, Maryland. Bob knew of my interest in the property because of my attempt ten years ago to gain permission for watermen to use the site. I worked with John Orme, then-president of the Anne Arundel County Watermen's Association, to make a deal with county officials and Johns Hopkins University, owners of the property, whereby watermen would be able to use the site for their boats and to haul and ship their harvests. That effort failed because the university and its Chesapeake Bay Institute wanted more money than the county could afford under its economic development plan. Hewison, however, obtained the vacant facility for an incredible price. His computer business occupies the largest of three buildings, some 400,000 square feet of space. The extra space, coupled with a new county administration, may make the original scheme feasible.

I agreed to meet Bob at his house, which is less than ten minutes from mine; it's about the same distance from his house to the Hopkins site in the opposite direction. I could feel the adrenaline flowing through my body when I pulled my truck into Bob's driveway. With Hewison, the watermen may just have the break they have been anticipating for more than a decade. Bob had just awoken from a short nap when I arrived. He runs crab pots and pound nets for rockfish, and purchases crabs from a number of crabbers to sell either wholesale or retail from his house. His daughter, Lorien, met me in the yard as Bob and I were getting in his truck. Lorien runs the retail crab and fish business during the summer and will do so for another few weeks before she packs up to begin her freshman year at Eckerd College.

While making the short drive to Shady Side we had an opportunity to talk about how we would approach developer Adam Hewison. Watermen by nature are skeptical, and both of us are skeptical of developers. We agreed that we would lay out our goals and see how he responded. We also agreed that as long as we got what we wanted, we didn't care if having watermen at the site would help him to develop "Discovery Village: where science and technology meet the Bay," an education center he wants to build with his wife, a public school teacher.

When we arrived at the large site situated on a point looking out over the Chesapeake, we were both excited and a little apprehensive. The three buildings looked deserted but there was a car in front of the main building. We decided to look inside and found the first floor of the building dark. Shadows made by sunlight streaming through the windows of the empty lab rooms scattered down the abandoned hallways. I told Bob I would search out the second floor because we were unsure where Adam was in this 400,000-square-foot structure. The second floor was no different from the first. I walked slowly through a large, dark room at the top of the stairs, guided only by a vague light shining under a door at the other end. As I opened the door I looked straight down a long hallway that was alive with light. I called to Bob, and by the time he made his way to me, I was looking into offices occupied by Adam's employees. A young man told us we would find Adam at the end of the corridor.

Adam Hewison has a British accent and a relaxed and pleasant demeanor. He immediately put us at ease, greeting us with enthusiasm and immediately offering us a cold soda; we chose orange-flavored bottled water and a tour of the facility. We would talk as we walked. Hewison explained that the second floor would house his computer business, while the twelve lab rooms, large assembly room, and other rooms on the first floor, would be designed and remodeled for Discovery Village. After touring the main building, moving through a giant maze of corridors and dark rooms, we exited into daylight on the water side of the property. The point had little water but about fifty yards down the bulkhead and opposite a second small building, water was approximately ten to thirteen feet deep and there was a thirty-foot pier stretching out into the creek.

We paused to let Bob sit on a piling; Adam took another, to discuss the possibilities. He was enthusiastic about having watermen tie up in the six or seven existing slips and mentioned plans to expand the slip capacity. Bob nodded to me and I took the lead telling Hewison that we were looking for a place not only to tie up, but also to repair boats and engines, to unload the catch, to sell it retail on the site, and to truck it to the men's wholesale markets. He immediately agreed and went one step further. He suggested that the second building be used as a crab house where people arriving on boats or by automobile could sit and eat crabs in a place overlooking the water. He also said that we should look in the third building, as it had a large area with more than one block and tackle that could be used to haul and repair engines. It also contained an area where seafood could be stored live and moved out huge garage doors right onto the trucks. There was also a cement trough, perhaps sixty feet long, that Bob saw as a place to keep live catfish, instead of the live boxes he now uses out in the river.

Touring this third building set our hearts and minds racing. It was in great shape with cinderblock walls, sprinklers, water, and electrical outlets everywhere imaginable. There was PVC piping that looked brand new that could carry water to and from the creek, a large room for engine repairs, new block and tackle, and countless possibilities for a huge seafood wholesale and retail operation. We were thrilled at the potential outcome this meeting could have on the use of this facility. Adam was just as excited as we went from one area in the building to the next.

Since we realized we were all in agreement, we decided it was time to end the visit on a high note. As we left what would become the watermen's building, we passed a fourth building that Adam said was going to be demolished. He also said he had just put in a new state-of-the-art septic system "that would accommodate the entire population of Shady Side," and he was going to put in a new road to wrap around the buildings along the waterfront. I asked what the next step would be, and he told me that on August 3, Anne Arundel County Executive Janet Owens was coming down for a private tour and discussion. We were pleased when Bob was invited to participate. After some incidental pleasantries and small talk, we shook hands and agreed to get together again after the county executive's visit so we could plan how to gain the support of other politicians and state officials, including the governor and U.S. congressmen. We also concluded that this facility, if developed the way we all dreamed, would be a model for the East Coast and probably the most unique private educational facility in the country that incorporated working watermen and their businesses. We felt an excitement that Bob and I intended to keep throughout the ride home.

In the truck, Bob and I discussed what we had seen and our individual impressions of Adam. We were impressed with everything and saw endless possibilities. We also agreed that we would try to tell no one else about this venture over the coming months except Larry Simns, president of the Maryland Watermen's Association. We were concerned about the fragility of the project, and we wanted to make sure that the venture was not discredited before it could get off the ground. We were not willing to risk such a potentially beneficial project.

When we arrived back at Bob's house the topic had shifted to the court case of our mutual friend, waterman Danny Beck. The case would probably result in jail time and a fine for Danny, which was very upsetting to Bob. We also talked about the disastrous crab situation and Bob said that he had "five hundred pots in the water, plus buying them from other men," and hadn't sold a wholesale crab in two weeks. As we moved into his house he emphasized that, this time, it wasn't a matter of moving your rig to another location: "It was dead everywhere."

After finishing a glass of iced tea in his kitchen and talking some more about Danny I told Bob I had to get home for dinner. He stopped me midsentence, insisting that I take some croaker fillets or crabs home to Cindy. I told him I didn't want to take his meal ticket, but he insisted. We settled on a dozen or so large male crabs and were both satisfied with the exchange: my time this afternoon for his crabs. Forty-five minutes later Cindy and I were seated at our kitchen table looking out over the Bay, eating hard crabs, sweet corn, and sliced tomatoes. It was a fine way to end the day.

As Kenny Keen, president of the Calvert County Watermen's Association, Larry Simns, and others have said for years, the more development and recreational boating that occurs, the more pressure there is on watermen who are trying to carry on their way of life. The state of Maryland and its waterfront counties deserve much credit for diligently pursuing avenues that work for both the local communities and the watermen. The outcome of Adam Hewison's Discovery Village, with a section of the project carved out for the watermen, is still under discussion. However, it is my sincere hope that the general public will be able to read about the success of this endeavor in the local media in the near future.

RECREATIONAL FISHING'S IMPACT ON WATERMEN

Unlike crab statistics, recreational fishing numbers are easier to estimate. The Marine Recreational Fisheries Statistics Survey (MRFSS) was conducted in 1998 for the National Marine Fisheries Service of the U.S. Department of Commerce, Regional Fishery Management Councils, Interstate Marine Fisheries Commissions, and state agencies to consider the impact and importance of marine angling in fishery management decisions. The strength of the MRFSS relies on the two principles of consistency and comparability. By collecting nationwide data in the same way since 1979, the agencies using the survey can identify notable changes in recreational catch and effort trends and evaluate the long-term implications of management measures.

Using the MRFSS, one can paint a picture of the impact of recreational fishing on the watermen and on their highly regulated industry. Consider, for example, the comparison between recreational and commercial harvests in the Chesapeake Bay in 1998 (*harvest* refers to any fish not released alive, including fish used for bait or released dead; *catch* refers to the sum of all harvested fish and those fish released alive):

	Recreational	*Commercial*
Striped bass	13,463,000 lbs	6,715,000 lbs
Bluefish	12,778,000	8,299,000
King and cero mackerel	8,721,000	4,881,000
Summer flounder	12,523,000	15,170,000
Atlantic croaker	8,213,000	25,304,000

In Maryland, recreational anglers harvested 392,000 pounds of striped bass; 1,692,000 pounds of white perch; 1,126,000 pounds of Atlantic croaker; and 206,000 pounds of summer flounder. In Virginia, anglers harvested 294,000 pounds of striped bass; 1,165,000 pounds of summer flounder; and 6,730,000 pounds of Atlantic croaker.

Recreational angling has a definite impact on the livelihood of the commercial watermen but it also contributes significantly to the economic, employment, and tax base of state and local communities in dimensions that far exceed the effect of the commercial watermen. The increasing importance of recreational anglers and the growing environmentally conscious voting community have influenced legislators and regulators, frequently to the disadvantage of the watermen. This may eventually cause the watermen to be defeated by more powerful groups fighting for government support, resulting in the decline of the historic and treasured commercial fishing industry.

LITTLE KNOWN FACTS ABOUT THE BAY

Elissa Leibowitz, a reporter with the *Washington Post*, compiled a list of fifteen things many people may not know about the Chesapeake Bay. Consider these facts for your next trivia game:

- At 200 miles long, up to 25 miles wide, and 4,400 square miles, the Chesapeake Bay is the largest estuary in the United States and four times larger than the state of Maryland.
- The name Chesapeake is derived from the Native American word *Tschiswapeki*, meaning "great shellfish bay."
- The Bay holds about 19 trillion gallons of water.
- The population of the Chesapeake Bay watershed, which stretches over six states, is 15.5 million.
- The average depth of the Bay is 21 feet. Bloody Point, in the upper Bay near the Bay Bridge, is the deepest part at 174 feet.
- More than 3,000 species of plants and animals, including 295 types of fish, call the Bay home.
- Forty-eight rivers and 100 small tributaries flow into the Bay.
- About 50 percent of the Bay's freshwater comes from one source, the Susquehanna River.
- Sea level in the Bay rises about 4 millimeters a year, or 1.3 feet a century.
- Historians debate whether the Vikings in the eleventh century, the Italians in the sixteenth century, or the Spanish in the sixteenth century were the first in the Bay.

- Early names for the Bay were "Great Waters," "Mother of Waters," and "Great Shellfish Bay."
- The Bay is the biggest producer of crabs in the country.
- The most photographed lighthouse on the Chesapeake is Thomas Point Shoal Lighthouse, built in 1875.
- Top sailing destinations in the Bay are Baltimore, Solomons, Annapolis, and St. Michaels.

CHAPTER 13

Other Bay Harvests

While people throughout the region have crabs on their minds during the summer of 2000, pound netters and other fishermen are watching the Tidewater Administration's program to introduce croaker (hard-heads) into Canadian and Midwest markets. In addition, the export market for striped bass will intensify as the commercial harvest gets underway. Aquaculture, or farm raising seafood, is underway with indoor production systems working from tilapia to yellow perch and hybrid bluegill for the Canadian processing market. Interest in culturing red drum, flounder, and sea bass for the live market is increasing, and fresh water shrimp—prawns—are also being studied. Hard clam farms in Chincoteague Bay are in production, and farm-raised catfish have thrived. For instance, those processed in February 2000 totaled 50.9 million pounds at 78.4 cents per pound, and have been more productive and financially rewarding than they were in February 1999.

As coastal states move closer to adopting a new management plan for Atlantic menhaden, a new assessment continues to offer a mixed outlook on the health of the stock. Recently analyzed figures from catch data for 1999 indicate that the spawning stock—an estimate of the adult population—has fallen sharply for the second straight year.

Doug Vaughan, a National Marine Fisheries Service biologist who makes the annual menhaden assessment, said that the spawning stock was 32,800 metric tons last year, well below the average 40,000 metric tons and less than half the spawning stock of 87,000 metric tons in 1997.

Interestingly, as the spawning stock has declined sharply, the preliminary estimate of young fish that joined the coastal menhaden stock shows an increase from last year. Vaughan estimated that 2.7 billion menhaden were recruited into the coastal stock in 1999. If that number holds, it will be the first time since 1995 that the number of young fish would be above the key threshold of two billion, meaning that the population would have surpassed the minimum number of fish considered to be a healthy stock. This year's harvest and that of next year will give a better picture of the true numbers. For unknown reasons, the menhaden stock has suffered from a series of years of poor reproduction, raising concerns about the health of the stock and its impact on other fish that rely on menhaden for food.

Shad, on the other hand, swarmed back to the Susquehanna in record numbers in the spring and for the first time in nearly a century found almost the entire river open for their spawning run. Maryland and Virginia also reported strong runs of shad as a result of their restoration efforts. On the Susquehanna, for example, 130,000 American shad had moved through the fish lift at the Conowingo Dam by mid-May, and the spawning run was still in progress.

"We definitely have our new record solidly in our hands," said Richard St. Pierre, Susquehanna River coordinator for the U.S. Fish and Wildlife Service. "In one day alone this spring, 22,000 shad were lifted over the 100-foot high Conowingo Dam."

"Bringing shad and herring back to our rivers is one of the best ways to repay the farmers, the cities, the industries, and others upstream for their efforts to reduce the oversupply of nutrients and sediments washing into the Bay," said Bill Matuszeski, director of EPA's Chesapeake Bay Program.

This is a remarkable comeback considering Maryland closed its shad fishery in 1980 and Virginia followed suit in 1994. Maryland's stocking efforts in the Patuxent and Choptank Rivers began a few years ago. Now, American shad and its smaller cousin, the hickory shad, which take four to five years to mature, are returning to the rivers to spawn.

Shad may be the star of the Bay when officials talk about restoring migratory fish to the tributaries because it has had the most successful recovery since the return of the rockfish, but the real beneficiaries could be smaller fish: the alewife and blueback herring. Like shad, they spend most of their lives swimming along the coast but return to their natal rivers and streams to spawn, sometimes moving into much smaller streams and even ponds. Their return is vitally important to the health of the ecosystem because the strength and growth of predator species depend on the bait fish. A 1991 Bay Program report stated, "Of all the anadromous fish species harvested in the Bay, the river herring [alewife and blueback] experienced the most drastic decline in commercial landings." In 1931, more than 25 million pounds of river herring were harvested in the Bay, making them second in quantity and fifth in value of all Chesapeake finfish. By the 1990s, the commercial catch was almost nonexistent and only 1.4 million pounds were caught along the entire East Coast.

While watermen are often criticized for overharvesting, there is a difference between the small family-oriented business we see on the Chesapeake Bay and its tributaries and the offshore factory boats that take vast amounts of fish either from miles of longline or from huge nets. For the river herrings much of the population collapse was blamed on foreign fishing fleets. In 1969 alone, the foreign fishery is estimated to have taken 74 million pounds of herring on top of the U.S. harvest. Many fish were taken before they had a chance to spawn, thus sending reproduction numbers spiraling downward.

Any recovery, as we have seen and will continue to see in the migratory species, is hindered by numerous problems related to loss of essential spawning and nursery habitat—problems resulting from water pollution, dams, and other fish blockages mostly caused by man. Further, river herrings were often forgotten, destined for fish meal, pet food, or bait—not exactly a glamorous epitaph. Their larger cousins the shad, 35 million of which were released in nine rivers in Maryland and Virginia in 2000, are harvested for their meat, their eggs (roe), and their fighting ability on the end of a recreational fisherman's line. Without the huge amount of recreational fishing money coming in for the species, scientists have a difficult time justifying spending a lot of money on hatchery operations.

In 2000, however, the stocking effort was stepped up for the herring: 2.6 million hatchery-reared larvae were released as part of a multimillion dollar fish passage and restoration program funded with mitigation money from the construction of the new Woodrow Wilson Bridge over the Potomac. Having a recovered herring stock would provide more forage for predators in the rivers and the Bay. They will also be a source for nutrients in the smaller streams. And, while not overly marketable in the Bay region, they are popular, especially in New England, both pickled and salted. The herring will benefit from restoration efforts for the shad. By piggybacking on their cousin, in due time, they may just return to our region in large numbers.

In early June 2000, carp, those overgrown goldfish that search for dinner in the mud of upper Chesapeake Bay tributaries, were found to be dying off in unprecedented numbers in creeks off the Bush, Bird, Middle, Elk, Sassafras, and Choptank Rivers. Scientists blame a combination of water temperatures that seesaw, a bacteria they have not identified, and spawning stress. In an energetic ritual, male carp ram females to shake loose eggs, then fertilize them as they settle to the bottom. While it is not unusual to see a hundred or so dead carp worn out from the mating ritual floating on the surface or lying on beaches in the spring, in 2000 the number of dead carp had been as high as five thousand or more.

Carp are popular as a food source in Europe, but even though they have been prevalent here since the nineteenth century, they have never really caught on for any American's menu. According to Kent Poukish of the Maryland Department of the Environment, who is tracking the carp, "There are too many carp for the water available and they have outgrown their spawning territory. The species is very prolific. It's wide-spread and it is not kept in check by natural predators. We know that Mother Nature will do this kind of thing on a cyclical scale."

As we have seen with almost every species on the Chesapeake, carp are subject to a wide variety of life-threatening natural and manmade occurrences, such as water temperature fluctuations, rise or fall in salinity, storms, point and nonpoint source pollution, overharvesting, and more. "This die-off clearly demonstrates that nature throws curve balls at creatures in the Bay," said Mike Hirschfield, a vice president for the Chesapeake Bay Foundation, "but, then, so do we."

CLAMMING WITH HARVEY BOWERS

A decade ago watermen who chose clamming over oystering or crabbing could make a good living. It wasn't as strong in 1989 as in 1969 when a record harvest of 659,000 bushels was taken from the Bay, but it was good, and fifteen times better than in 1999. At the close of the millennium, clam harvests plummeted to 12,300 bushels and watermen were running from the fishery in search of better income elsewhere. Clam beds, which frequently lie adjacent to oyster beds, are particularly sensitive to nature's elements. Variances in water temperature or salinity can wipe out a harvest.

Then, in June 2000 the worst possible news hit the commercial clammer: A parasite, *Perkinsus Chesapeaki*, closely related to the Dermo parasite that has plagued Bay oysters for two decades and decimated much of the oyster industry, may be present in soft-shell clams.

Researchers from the National Oceanic and Atmospheric Administration (NOAA), the Virginia Institute of Marine Science, and the U.S. Department of Agriculture's Agricultural Research Service found the organisms in clams taken from ten Maryland sites in the upper Bay between 1990 and 1999. According to NOAA's Shawn McLaughlin, "It's too early to tell whether it causes a severe impact on the clam population, but in clams with advanced infection, it may interfere with reproduction and digestion because large numbers of parasites may appear in these organs."

Harvey Bowers picks clams off the conveyer belt and drops them in a five-gallon bucket. This skill requires tremendous hand-eye coordination and speed. The belt must keep moving to save time, but every clam that goes back overboard represents lost income.

My old friend Ann Pesiri Swanson, executive director of the Chesapeake Bay Commission and a Bay leader sympathetic to both the environment and the watermen, said, "We don't know whether it's related to the clams but there is substantial change in the harvest over the past decade. It does say that given our experience with the oyster diseases MSX and Dermo, we would be foolish to ignore it."

When I call Harvey Bowers, thirty-five, a native of Rock Hall now living in Chestertown, in August about going out clamming, he says he only goes about every other day since that is all the market will bear. Crabs and eels are weak right now, so there is no market for his razor clams, which are used for bait—unlike manoes (steamers or piss clams, as they are often called), that are for human consumption. We decide to meet at 5:30 A.M. on August 31 at his boat, which is tied up at the Talbot County dock next to the Newcum/Royal Oak bridge just outside of St. Michaels. He has a market for five bushels, or a $100 day, not a very encouraging prospect considering there is a one-hundred-bushel limit on razor clams.

When I arrive at his boat, *Busted and Disgusted,* he is checking the oil in the diesel engine that powers the hydraulics for the clam rig. It is pitch dark with overcast skies. Lights are visible from two other boats belonging to Captain Eddie Higgins and Captain Danny Ashley as they leave the dock and move out into the Miles River for the clamming beds.

We quickly follow suit in order to get to the clam bed Bowers wants to work today. Regulations state that a clammer cannot start work until thirty minutes before sunup and must be finished by 1:00 P.M. On the way out of the harbor he asks me if I have the time and I say, "It's about 5:35."

"Nah, I need it exact. I'll get it," as he reaches for his radio and calls for the time from one of the other men already out of sight. "What time you got?" he asks.

"My timer is busted. Danny can give it to you." With that direction, a voice comes over the radio saying it's 5:41.

Bowers plugs the time into his loran and marks his coordinates, which show me we have 1.11 miles to go to the specific site he wants to work. "Sun's up at 6:14 and we can go in the water a half hour before then," he said. "It's a $260 fine if you violate it on either end."

Riding out to Gibson Flats in the Miles River, he says that all the rain we've had this summer hasn't affected the clams, which is very unusual. "Damn if I can understand it. Usually low salinity will wipe us out and vice versa.

"We can clam year 'round and I do except for some tonging oysters for a few months at the beginning of the season. Half the time you can't figure out what's worse, the oysters or the clams. Our biggest time of the year is the summer when kids are out of school. At the resorts, especially up north, people eat fried clams like we eat french fries so the summer is hot for us in a lot of ways. A bushel of manoes can go for $110, but we ain't got any to sell.

"With all the fluctuations and frustrations working all over the Bay from the Virginia line on up, and a wife at home, I got out for a couple of years. I worked off the water and didn't like it so I came back. This is what I love but really it's all I know. I was just burned out a few years back and took a break. I know I'm in it for the long haul now but it's hard on you. All this work is hard on you. I've got a bad neck and back. Just over ten years ago I went to the doctor and he asked me how old I was. I told him twenty-three and he said, 'Well, you might be twenty-three, but your back's forty-three!' I've been in car wrecks and that coupled with workin' out here does a job on you. Why one time I was wearin' so many braces and casts the only thing I *wasn't* wearin' was a halo! Don't talk to me about bad backs. I don't want to hear about that from anybody."

I ask him about his major concerns and about the environment.

"It ain't the environmentalists that hurt us so much. Oh, they raise a lot of hell and I'm sure they hate clammers as much as the oystermen, crabbers, homeowners, and everybody else does. Fact, we're probably the most hated ones out here on this water. It's the regulators and the concerns over what we're doin' to the SAVs [submerged aquatic vegetation] that bother me the most. Right here on this river, and this is a hell of a river, we worked that shoreline over there hard this spring. In three months, up to now, there's little clams all over that shoreline and the rest of the river too. That's a good sign. There never used to be clams here.

"Here's an example of a big issue I'm fightin' right now and this is what it's all about. The state contracts with VIMS [Virginia Institute of Marine Science] to do a grass survey of the river. They send this fellow Bob Orth up to do aerial photography/charts to show where the grasses are and send him underwater too. Now they do the survey every July and it shows there's a lot of submerged grasses and then the state closes those areas to clamming.

"Well, I'll tell you, there ain't enough grass here to put in a rollin' paper. That whole shoreline over there," as he points east, "is closed, and

A clam dredge with its long conveyer belt is stored parallel to the cabin. The vacuum rests on top of the rig until it is placed in the water. Then the hose will blow the sand and mud off the clams as the end of the belt lying on the bottom carries clams up to the clammer's hands.

there's not a blade of grass to be seen. I got a citation for clamming in SAVs and I'm fighting it. It's not the money; it's the principle. VIMS says one thing; Maryland says another. We're caught in the middle. I want more frequent aerial surveys, not just once a year. How can we expect to make a livin' when the rules are a mile thick and they change them all the time?

"And I'll tell you another thing. Why would I want to clam in a grass bed when it clogs up all my hoses, my pumps, and everything else? I've got until one o'clock to catch my limit or make market and they think I want to spend all mornin' cleanin' out grass!

"I even went and got a lawyer. He told me he never did an SAV case before and he would charge me about $350. Then he said to do it right he'd hire our own biologist to go to the clam bed and see if there was grass there. That would be another $1,500. I talked it over with Simns and he said that if we send our biologist in the water and he sees some little sprouts comin' out of the bottom, I'm screwed. I said the hell with it. But I'm still fightin' these closures based on Orth's word.

"I know where the clam beds are by followin' the bottom: hard, sandy, mud, shells, stones. I like sandy bottoms because they aren't hard on the rigs. I'll find the clams if I can just get to a place where the state doesn't say there's grass there. Take Hunting Creek over there. It was closed forever. We talked to the state and they said they would open part of it up to clamming. Just as soon as they did, they reversed themselves and closed it right back down because they said there were SAVs up there. And, I'm gonna tell you like it is, Cap'. Like I said before, there ain't enough grass here to put in a rollin' paper.

"One of my complaints with the Watermen's Association is Simns gives up too soon. He plays the games of politics all the time but, to me, he's too quick to negotiate. He'd say, 'Take this part away from us but not the whole river.' Well, we don't agree with that. We don't want 'em takin' any of the rivers unless they can prove it to us that there's a reason. I know he's got a tough job but sometimes you've got to say knock it off. He don't like to do that 'cause of all the issues out there and he hopes to trust the state. If I light a fire at that end of the point pretty soon the whole peninsula is going to burn up. That's what will happen here. You give an inch and they'll take a mile.

"And, when the state does something to shut down the crab industry, which they will do sooner or later, it will have a huge impact on clammers, gear manufacturers, picking houses and those ladies who work there, fuel tax income—there will be hell to pay then. We all feed off the crab harvest."

As we reach the clam bed he puts the boat in neutral and moves from the cabin, which is hot and humid because its windows don't open, and moves to the stern section to get ready to go to work.

First he turns on the auxiliary diesel engine that sits on his deck. This engine will run all the hydraulics involved in the process. He unties the forty-foot long conveyer belt, made of stainless steel links, and drops a six-inch water hose down in front of the conveyor. This hose will take water from a pump attached to the engine and blow clear the sandy bottom as the conveyor moves forward to bring up clams, horseshoe crabs, shells, sticks, rocks, and a few bottles. He sets three five-gallon plastic buckets on the washboard or gunnel between his body and the conveyor belt. Bushel baskets are set on the boat's engine cover. With everything set up and the conveyor and hose in the water, he's ready to start the belt and bring up the clams. It is about sunup; I see him scan the area, and I wonder if he's looking for police who may think he's a little early. As he stands with the buckets in front of him and parallel to the side of the boat, the boat's engine throttle and rudder stick are to his left. Directly in front of him at chest level is another engine throttle, used to raise or lower the conveyor to match the distance to the bottom. Across the conveyor is another gear shift to stop, speed up, or slow down the movement of the belt. Directly behind him, just above his head, is a radio, not for talking to other boats but for listening to country music, Mary Chapin Carpenter at this point in time.

As the clams and miscellaneous debris move up the belt past him and drop off the stern, he moves like a Detroit assembly line worker, his dexterity and speed unbelievable, picking clams from the belt and tossing them in the bucket. We are working about 150 feet from shore and throughout the morning the boat will move in circles with about a quarter of a mile radius. We don't cover a wide area, as this bed is small but productive. As I watch him work I am amazed that very few clams get past him.

As the sun fights to break through the thick cloud cover, Bowers stops and asks me to take pictures of Danny Ashley's *Tammy Lynn*. I readily agree and go to the cabin to get the camera.

"Danny taught me clammin' when I was a teenager. He's a hell of a waterman. Used to hand tong oysters barefoot. He'd stand on the washboard and work those shafts with his feet like you wouldn't believe. Never seen anything like it," he said.

Without thinking he goes back to his four different gears and the radio volume control. All his actions seem to be instinct—reading the bottom and pulling the clams off the belt. Efficiency is everything when the clock is running and time is money.

As he works, he explains that the clam rig cost him over $10,000 in used equipment: $4,000 for the diesel engine, $400 for the motor to run the conveyor belt, $200 for conveyor chain, $150 for cable to raise and lower the conveyor, $5,200 for a new pump to move water from the port side hose through to the black hose on the starboard side at the front of the conveyor, and so on. "It's a damn shame, but you can buy a lot of good used gear nowadays 'cause so many men are getting out of clamming. There's money to be made but it's by fewer and fewer guys." I think to myself that this $100 day is not contributing much to help cover his expenses.

Harvey Bowers sums it up perfectly when he says with a smile, "There's nothing cheap on this boat but the labor!"

He pauses to point out a cluster of small clams on the belt to prove his earlier point. "There's no size limit to what we can catch but this is all I got. This is what I do. In the long haul I want this baby to grow up so I can catch him another day."

"How long will it take for this little one to get to be three inches long?" I ask.

"If we have a temperature drop it will sprout like a weed in the garden," he casually responds.

I move in and out of the cabin trading my camera for paper and pen. For the second time this morning he watches me bump my head coming through the cabin door. He smiles and says, "The boy that built that cabin built it for me," as he raises his hand up over his head to illustrate his height of about five feet, seven inches, "not for guys six feet tall. You'll bump your head a time or two again before we're through." I take his word for it and try to prepare my mind likewise (without success).

Harvey sells most of his clams these days to William Clark and Son Seafood in Chestertown: five bushels today, fifteen yesterday, and fifteen is the demand for tomorrow. Clark likes to keep them fresh so he only takes what he needs on a daily basis. Danny Ashley sells to Clark as well. Harvey has other buyers he can also count on in other parts of

Maryland and Virginia. "Yesterday I had my market by 9:00 A.M. and we'll do that today too. The clams are comin' and I want to get the hell out of here before it rains."

As the bushel baskets fill and he reaches his market for the day, he hoses off the belt, the washboard, the deck, his oilskins, and the engine cover. "These rigs are hard on a boat. They're worse than the Bay herself I believe," he says as I move to the cabin to get out of his way. He must raise and secure the conveyor and hoses and put away the buckets and baskets.

Once everything is washed and secure he settles into the cabin with me and we head back to the dock. We talk about politics, about the Watermen's Association, about Danny Beck's getting prison time for his rockfish problem (Harvey is a big supporter of Beck and hopes that he will become president of the MWA when Larry Simns retires), and about other topics of his choice. He mentions the movie *The Perfect Storm* and says how much he liked it. I tell him I saw it and enjoyed it as well, and I suggest he read Linda Greenlaw's *The Hungry Ocean*. She was out with the fleet when *Andrea Gail* went down and wrote a book about a month on a swordfishing trip.

"I'll tell you what. I'm not scared of anything out here in this Bay but I'll be damned if I'd go out in any ocean with those huge waves crashin' all around me. No sir, Cap', I ain't goin' out in any ocean water."

I tell him that this seems to have been a good day, not because of the money, but because we were out and back in short order and the clams were in good shape.

"Well, yeah, but we've got to get more clams out here and we've got to get the state off our back. I know I'll get off the water someday but I don't know anything else so I can't leave her yet. The state with all its regulations can bury us no matter how much we do to try and help the Bay, watch what we harvest, and everything else. It's like we don't have enough to worry about just wondering if there are going to be enough clams, oysters, crabs, or fish to meet the market demands when we have a bad run. Like the manoes, we could make some serious money but they're aren't any around now."

Eddie Higgins comes over to joke around and help Harvey take the clams from the boat to the truck. His arms are as big as my thighs as he lifts the bushel baskets onto the truck bed with ease. "Come on, Cap', I got to get out of here," he says, hustling Harvey along.

As we are talking, an older man arrives to meet the men. They greet each other and I am told by Eddie, "Here's an old-timer. Good water-man you can interview for your book. Hell, he even knows how to tell if you're goin' to have a heart attack!"

"Damn right," he says. "Look at your earlobe. If you got a crease in it you're goin' to have one."

Harvey steps up toward the bulkhead and asks, "How's mine look?" with a smile.

"Clean as a whistle. Ain't no heart attack in your future."

"What about mine?" Eddie follows with a smile as we all look at a huge indentation in his earlobe.

"You're good as dead now," the old gentleman says seriously as we laugh.

Eddie says he can't listen anymore to his fate. The older fellow leaves and Danny pulls away in his truck.

Harvey and I part company, so that he can take his clams and Captain Ashley's up to Chestertown and so that I can drive back to Fairhaven. I offer him a copy of my book *Sunup to Sundown* and he gives me a shirt from the Rock Hall Watermen's Festival with an invitation to attend next July, as well as an open invitation to go out with him again anytime.

This was a short but interesting morning with young Harvey Bowers, a waterman all his life and what he'll always be.

CLAMMING WITH BILL CROOK

As I was working with Harvey, *National Fisherman* reporter Elizabeth Scott Leik was spending time with Bill Crook, who catches soft clams for human consumption. She obtained another perspective from a man who has clammed for more than a decade. Crook works out of Whitehall Creek, a couple of miles north of Annapolis. The clam beds he will work in the rain are across the Bay off Kent Island. Unlike Harvey, he will have a maximum eight-bushel limit.

An hour later, *Leslie,* with Crook and Leik aboard, arrives at a party consisting of five other clammers setting up their rigs just off Bloody Point. As he sets up he spots *The Lady Kate.* "That's a wooden boat," he says. "I know because that fellow had a fiberglass boat and went back to wood. Few people make wooden boats anymore."

Leslie, named after Crook's wife, is his third boat. It is also wood and was built by Jack Willing on Deal Island using sturdy, mahogany-on-oak construction. Most wooden boats on the Bay are pine, cedar, or oak but Crook prefers mahogany. "I like the feel of a wooden boat. It's more stable and I'm more confident in it," he says.

As he sets his anchor in ten feet of water about a mile from shore, he explains that clamming is usually a one-man operation. There are a couple of days a year he needs help, and on those days, he takes along his thirteen-year-old son Nicholas.

Like Bowers and the other clammers, Crook uses a well-worn hydraulic clamming dredge. On the port side of *Leslie* sits his six-cylinder Cummins auxiliary engine for the conveyor; its exhaust creates the vacuum that primes the dredging rig. He connects the suction hose, which also runs across the deck from the Cummins, to a manifold at the head of the rig. With it in the water and on the bottom, jets of water bore into the mud, clay, or sand to unearth clams. The large conveyor belt slips into the water, and at the front of it, jets blast twelve to eighteen inches into the bottom blowing rocks, old shells, bottles, and clams onto the belt. The process is identical to that of other clammers. The only difference is the type of clam they are harvesting.

Crook has been clamming for the past twelve years. He can generally find enough of the soft shells to supply out-of-state markets as well as local restaurants. He ships the majority of his clams to Seabrook Shellfish in New Hampshire. "That came about by me just looking around while I was on vacation in New England. I gave my card to a local guy and he passed it along, and I got a call from Seabrook a few weeks later. I hadn't even met the guy I was going to sell to; it just kind of happened that way. We met after I had been shipping to him for a while," he says.

At the end of the day, Crook has six bushels in his cooler. "That will make a packed five, when each one is filled to the top," he says. By mid-May the catch can increase to fifteen bushels.

I remember a decade ago when clammers had a poster that read "Catch eleven and get paid for ten" because in order to make decent money equivalent to ten bushels they had to catch an extra bushel.

Nonetheless, clams have dropped off since that time in 1989, when 365,000 bushels with a value of $10 million were harvested. In 1991 the total catch dropped to 182,000 bushels, and a severe heat wave in 1992 cut the harvest to an unbelievable 19,000 bushels.

In 1999, 12,500 bushels were harvested with a value of $1 million. "Even since 'seventy-two, Hurricane Agnes, the clam harvest has never been the same; it's been an unstable business," Crook says. "We usually don't have enough clams by August."

Maryland is the southernmost area where clamming is still a commercial business. Clams are found in Virginia and farther south, but their short life expectancy doesn't support commercial watermen. And, as clammers know as well as any other watermen, nature takes her course in other ways, as well. "The drought brought more cow-nosed rays in the Bay to spawn and for the past ten years, they have been staying longer," Crook says. "Those rays eat clams by the dozens."

Crook tells Leik, "The strangest thing I ever dug up while clamming was a set of false teeth with a gold tooth in front. I didn't really understand that, but I kept it." It sits in his cabin behind his seat.

Bill Crook also fishes for crabs or oysters. There used to be a natural season for clams inspired by weather. "When it was warm, you came out, usually April to December maybe. You could catch what you wanted to." This year clamming is open from early morning until 1:00 P.M. from April 15 until September 30.

In 1999, there were 163 clammers working the Bay in Maryland for the 12,500 bushels of soft shells that were caught that year. This number does not include the razor clams that Harvey Bowers harvests to sell for bait. Soft clams for human consumption must be two inches in size and boats must be equipped with a sixty-degree cooling unit for storage. There remains a moratorium on clamming licenses (a commercial fishing license) and a waiting list to get one in Maryland.

With a capital investment of up to $100,000 for all new equipment (or as Harvey Bowers said, a used boat for about $30,000 and a hydraulic dredging rig for $5,000 or more) the clammers want to get their $70-to-$100-a-bushel fee for the clams because of the catch limits. Even at that price, Bill Crook and Harvey Bowers, who harvest different types of clams, did not break any records with their income from the clams.

EELS: AN ALTERNATIVE FISHERY

Most people don't think of the eel as the most attractive species of fish living in the Chesapeake Bay. It is not as appealing as the rockfish and, for most, it is an annoyance when it swallows a hook, even though a large eel can put up a pretty good fight.

I remember catching them as a youth while perch fishing from our pier in St. Mary's County. We loved the battle but had a difficult time with the eel's slimy skin when trying to remove the hook. To a ten-year-old, a nice-sized eel of twelve inches or more is a rather frightening catch. We would end up taking out our frustrations on the aggressive fish, I regret to say, by clubbing it to death just to free ourselves of the problems it caused. I know there are folks who enjoy smoked eel but I personally have no desire to partake.

The diverse opinions about eels remind me of a time years ago when I told SeaTow owner-operator Dave DuVall of Annapolis (but a native of the watermen's village of Rock Hall) that I was going over to the Eastern Shore to get some piss clams (soft-shell clams or steamers). He looked at me in amazement and said something to the effect of, "Don't tell me you're going to eat them! I'd never use them for anything but bait!" I guess it all depends on your perspective, and I know what my perspective is on the eel.

Despite the diverse opinions, eels can provide a good financial resource for watermen who choose to catch them, regardless of the fact

An eeler, one of only a couple of dozen on the Bay, empties his catch from the pot to a barrel. Eels can bring up to $4.00 a pound for export to Europe and Asia, but the market is unreliable—when it's good, it's very good, but when it's bad, it's very bad. Courtesy Maryland Watermen's Association.

that eelers always deal in an unstable market. In June 1999, Bill Legg, a respectable waterman in his own right, was running about nine hundred eel pots and making an average of $1.80 a pound. Eels are considered a delicacy in Europe and Asia and can provide an alternative in a slow-moving crab season.

An eel pot is a cylinder about eighteen inches long and eight inches across. Eel bait is razor clams—a bushel a day, costing about $25. The bait is put into the pot, and one end is tied while the other is left open so the eel can get to the bait. The pots are hauled in much the same way as crab pots are, but the eels are dumped into a large tank instead of bushel baskets.

POUND NETTING WITH TOMMY HALLOCK

By mid-August crabs remained scarce regardless of whether men were trotlining or potting. Many began to augment their meager incomes by gillnetting or they got out of crabbing altogether. I thought that despite the flurry of rain and freshwater pouring into the Bay, pound netters might be doing something, so I decided to call Tommy Hallock, a fourth-generation waterman from Shady Side, Maryland, a small peninsula about twenty miles south of Annapolis. Hallock has a good reputation and could give as good a representation as anyone on what good fishermen were doing. We did the usual trading of telephone calls, and when we finally connected he was glad to hear from me and happy to take me out. We decided to meet "a little before 5:30 A.M." at his boat, *Miss Cindy,* on August 11, 2000.

Following his directions reminded me of the directions I got from Danny Beck and other watermen: "Come into Shady Side, turn past the ball field, and then make a right after the Salem Avery Museum. There's a sign that says 'Point Pleasant.' Pull in there and you'll see a house under construction. Go behind it and that's where the boat is. You'll find it, but in case you get lost here's my cell phone number." I always follow these directions with confidence that they will be as easy as they seem. I know better than to ask for more specifics or clarification because watermen always seem to expect to be understood the first time they explain something. Asking for further clarification seems to frustrate them. Unfortunately, the directions are never simple because they don't think about the other ball field, the left and right you have to take before finding the museum or other landmark, and other similar complications. After making several wrong turns and backtracking, I reach the boat a little before 5:30 A.M.

Hallock is a good-natured fellow in his midforties. He is strong and tan with blond hair. A week earlier he was in Jamaica on his honeymoon. Now he's commuting from his wife's home in Fairfax, Virginia, until they can buy another house in Shady Side. The house under construction in front of his dock is on ground that Hallock and his sister inherited when he was thirteen and she was fifteen. Their father had passed away, and when his sister didn't want to live in the family house, he had to sell. "I couldn't get money to buy her out at thirteen so what else could I do? I was raised on this point. We owned four acres. Now I'm trying to buy this house here that's been neglected for over a year and I'll have to pay more for it than they paid us for the whole four acres years ago. I hope it works out so I can live here and be with my boat and gear right where I started as a waterman when I was thirteen."

Standing beside his truck talking in the dark, I look across Parish Creek and notice lights in a building—most unusual in this area.

"Is that the Johns Hopkins property?" I ask him.

"Sure is. I heard you went over there to meet with Adam and scope things out a few weeks ago with Bobby Evans."

"Yeah, we had a good meeting. I liked Adam but I told Bob we need to ask for more than we think we might get. Then we'll be satisfied with the results. I think the county executive went down on the third to take the tour."

"She did," he said, "I haven't talked to Bobby. I only hope it all works out and that Adam doesn't fold up shop in a few years and leave us dangling out there like what always happens."

We talked about the creation of a watermen's landing and seafood operation, and I could tell that he, like so many others, had hopes for the future with this facility. The watermen have been used too many times and disappointed even more. I, too, hope it doesn't happen again.

As we walk down to the boat, I can tell he arrived well before me. A dozen plastic bushel baskets and another dozen fifty-pound plastic tubs were on board and his Marlboro Lights and coffee were in the cabin. He reached through the open starboard window and grabbed the cigarettes. He was content to stand on the dock and wait for his crew, and so was I. We talked about his crew: Charlie Quade from Churchton, whom I know and have seen often over at Bob Evans's house, had sold his crabbing gear and has partnered with Hallock so they can catch the quota of

Leroy Philip has crewed for Captain Tommy Hallock's family for over fifty years and still provides humor aboard the boat every day.

two licenses. Andy, a student at Anne Arundel Community College, is on his second season with *Miss Cindy,* and Leroy Philip, a black man of seventy-six years, started working with Hallock's grandfather on the water fifty years ago.

"You can start to talk to Leroy, but if he clams up then don't waste your time tryin' to get him to say anything. A while back we had a fellow on the boat and he tried to talk to him with no success at all. It depends on his mood. He turned that son of a gun off. If he does want to talk he'll tell you about ten of his stories that I have heard about 40,000 times," Hallock says. We both laugh at the thought as Charlie comes down the pier.

"Well, good afternoon," Tommy says with a grin.

"Same to ya," Charlie responds. "Hey, Mick, how you doin'?"

"I'm fine. How's your back?" I respond, recalling that he pulled it very badly while perch fishing in the winter.

"Pretty good. I wear this brace now and it helps."

"Bull." Tommy says. "He's as lazy as the rest of the crew." We all smile, knowing that Charlie Quade is a fine waterman and Hallock's partner.

We all climb aboard the boat. I store my duffel bag in the cabin as Charlie reaches into a bag he brought and pulls out a new pair of white oilskins for Tommy. As he hands them over, Tommy eyes the pure white coveralls and says, "These are so pretty I won't know how to work in 'em."

Andy had picked up Leroy at his home in Galesville and they are arriving as the clock strikes 6:00 A.M. We shake hands and make introductions as Tommy starts the engine. There is a nineteen-foot Sea Ox skiff tied off the port side of *Miss Cindy*. As we back out of the slip and turn to head out of Parish Creek into the Chesapeake, Charlie pulls the skiff and ties her off from the stern. We will tow her to the pound net up the Bay just north of the South River. Hallock has two pound nets, one a little south of us and one six miles north. He works them on alternate days. Today he wants to go north.

The gear on board for pound netting is relatively sparse: baskets and tubs, a large wooden culling board that stretches the width of the boat and rests on the engine cover amidships hanging slightly overboard on each end, a large net with a hoop about two feet in diameter with a homemade cedar handle about eight feet long. A line on a hydraulic winch runs off a boom overhead. Tommy will attach the line to the net when he lowers it into the pound net to catch fish.

Running out into the Bay the water is flat and calm as the sun appears like a large red balloon sneaking over the edge of the horizon. For the next couple of hours, as it rises in the sky as if tethered to an invisible string, it will strain unsuccessfully to burn off the haze that blankets the Bay with humidity. Charlie and Andy join Tommy in the cabin while I am left on deck with Leroy, wondering whether or not he will talk to the stranger on board this morning.

For a few minutes, as he ambles over and starts puttering with the bushel baskets, he is acting like a shy child who has something to say but is hesitant to make the first move.

"This rain has us playing hell with mosquitoes, doesn't it?" I ask, deciding to break the silence.

"Sure has. Why the other night I was fixin' chicken and opened up the kitchen door for a minute. I thought the mosquitoes were goin' to carry the chicken off into the woods. Swear to God," he said laughing.

For the next forty-five minutes I couldn't keep Leroy from talking. He spoke with pride of his two daughters who never smoked or drank, had professional jobs, and nagged him about quitting work. He told stories of bootlegging whiskey in the early fifties at baseball games that were played every Thursday, Friday, and Saturday night; of making fifty cents a day pulling grass from pound nets when he was thirteen years old and without a father to help support the family; of buying cigarettes at two for a penny; of making Hallock's pound nets, which he still does; of seeing Eastern Shore oystermen told to leave the oyster beds of Parish Creek by local watermen sitting on the bows of boats holding rifles to protect their beds; of serving in the Navy and coming out to work for Captain Bernard, Tommy's grandfather; and more. He would laugh with every story, and while I often had a difficult time understanding his words and dialect over the hum of the engine, I picked up enough that I could laugh with him. As is often the case with older folks, he had much homespun wisdom between the lines of his stories.

"Why, Captain Bernard and I would leave the creek and head for the James River or the Rappahannock and I would write our course down on a piece of paper as he would call out every marker or landmark, the time it took to get there from the one before, the speed, the wind, everything so that when we came back, if there was a fog, we wouldn't have no problems. In all them years comin' up or going down to Virginia to get seed oysters we never messed up once. Had it all down on paper. Yeah, we'd go down there and buy oysters for Woodfields or somebody else and bring them back up here for the boys to plant in the rivers. Plant them in the rivers so you have a bed to work when it's too rough out in the Bay because we hand tonged back then. Just me and him. We did it together. He was a kind man, just like Tommy's mother and father, and Tommy too. Good folks.

"Captain Bernard always trusted me with the money 'cause unlike most boys back then I could count and never tried to cheat him. I would go into the place in Virginia and give the man the money for the seed and come back to the boat and put the change on the table. Capt'n would look at me and say, 'That's your change,' and I'd say 'No, sir, Capt'n, this here's your money,' and he'd say, 'The hell it is. You gave the man money and he gave you change. It's your change.' Might be two, three hundred dollars and he'd make me keep it. He took good care of me and never would let me work for no money. Been right here with them for over fifty years now."

I asked him about the crabs being off and how the fishing has been.

"I know every inch of this Bay. You can't tell a thing about the crabs. I feel sorry for them boys. Now, the fish, they been good. There's more fish out there than you can count but it tears you up that you have to throw back a beautiful twelve-inch rock."

I hear the sound of the throttle pulling back as Andy and Charlie leave the cabin. Turning to look off the port side I see the forty-foot long pine poles sticking up out of the water fifty yards ahead. Gulls and cormorants are on each piling; dinner lies directly beneath them in a mesh trap. I imagine the hunting must be good this day. They quickly leave the pilings as we pull up alongside the outer poles and tie off *Miss Cindy.* Charlie tells me we are opposite the Arundel on the Bay community and Fishing Creek. Annapolis lies a couple of miles to the north.

The pound net begins from a point offshore with a thousand feet of net stretched out between poles. This is called the leader, and the fish see it and follow it to an open pen called the heart. From the heart there is a funnel into the pound, and from the pound they go through another funnel to the pocket, where they wait until they go to market. The poles are rough pine, and each season the men have to knock the knots off new ones to make them smooth so that the net doesn't get caught up on them.

The nets are handmade in the off-season between November and February and dipped in copper antifouling paint to prevent their decay. It takes two fifty-five-gallon drums of paint to coat each net at a cost of

Captain Tommy Hallock, *right,* and his crew from *Miss Cindy* gather a pound net to bunch the fish. When it's tight and the fish are close together, he will scoop them up with his net.

$800 a drum. Usually, the paint will last from April until August before the nets have to be redipped. This year, however, all the rain brought in more freshwater, which causes algae blooms, so the net was pulled and repainted in June. Charlie had driven the 260-mile round trip to Virginia to pick up two drums of paint yesterday because here in August, the $1,600 expense had to be made again. The net will be pulled and painted in the next week or so. The men hope for dry weather and a rise in salinity to keep the algae off the nets through the fall.

With the boat tied off parallel to the end of the pocket the three crewmembers get into the skiff, circle around to the other side of the pocket, lower it, and take the boat inside. Then they maneuver to the south side of the net and untie the lines from the poles. Tommy is at the stern of *Miss Cindy* untying lines from the north side of the pocket. With lines undone the three men begin to gather the net into the boat.

As the net comes in, Tommy loosens the ties to the bottom of the net so it can be pulled up, forcing the fish to come to the surface. Excitement among fish and men builds in anticipation of the next move. As the net is pulled into the boat, long metal rods are placed in the gathered section with both ends braced across the skiff against the inside of the boat. The rods keep the heavy net from falling back into the water as it is gathered.

With each pull, the pocket gets smaller and more and more fish are frantically swimming near the surface. Tommy takes a ten-foot two-by-eight and creates a makeshift bridge from *Miss Cindy* to the skiff. The men walk across and into the bigger boat setting up the culling board and baskets. Leroy stays in the skiff to finish gathering the net as fish are removed; Andy and Charlie are on one side of the culling board and Tommy is on the other. Tommy attaches the big dip net to the hydraulic line, swings it over the pocket, and lowers it with his foot on a

Tommy's stepson Jacob looks on as the crew pulls in the net. As it draws closer to the boat, fish migrate to the surface.

gearshift while he works it into position with help from Leroy. The net is lowered into the pocket and then raised, with about 350 pounds of flounder, trout, croaker, menhaden, rockfish, perch, and mackerel inside. Tommy swings it over the huge culling board and pulls a line to open the bottom of the net. Fish drop out on the board and on the floor, and some go back overboard. The men then move with incredible dexterity to separate the fish by size and species, putting them in different baskets. Charlie then pushes the bait fish and small rockfish overboard with a board the width of the culling board.

Tommy explains, "With crabs off so much there's no demand for the bait fish so I put that back overboard. We'll see a minimum of $700 worth of bait fish go in the water today. Damn shame."

Andy, Tommy, and Charlie are covered from head to toe in rubber as water, fish scales, and grime begin to fly everywhere with each opening of the net. I have moved, on good advice, to the roof of the cabin to take a dryer point of view.

When the last fish is taken from this set, the net is dropped overboard; the crew get back in the skiff and move to the southern side of the pocket. They start the process of gathering the net again. Tommy decides that the tide is moving so fast this morning that it will take two turns on the pocket to draw it up tight enough to clean it out.

The catch today will bring about $2.00 a pound for flounder, $2.00 for rockfish, $1.25 for trout, $1.00 for perch, and $.40 for croaker. It was a "piss poor" day with about ten thousand pounds of fish brought in, counting the bait fish, worth about $800. For the men, an average good day of fishing should yield about $2,500; on a rare day when they bring in forty thousand pounds of fish with high market prices they might make $10,000. Then again, it's not unusual to make far less than $800, and that doesn't go very far among four men, a boat, gas, and gear.

With only ten thousand pounds of fish coming from the pocket it was a quick morning. On the way back to Parish Creek we had an opportunity to talk.

"This was a damn poor day," Hallock explains. "We're on a full moon and we catch more fish on the dark side. I'll tell you every day

Tommy Hallock prepares to
empty the large dip net,
loaded with fish, onto the
culling board.

changes because there are so many variables: tide, salinity, temperature,
the moon. One day five thousand pounds and the next day thirty thousand pounds. The only thing that's constant is the regulations. We can
pound net in an open season but rockfish, for example, has a
forty-eight-hundred-pound annual quota and that's with both my and
Charlie's licenses together. The most we can bring in is sixteen hundred
pounds a day and we have to tag every rock, so let's say in September we
reach our forty-eight hundred pounds for the year—every rockfish after
that gets thrown overboard. More fish die of old age than we can catch.
But the fish have been unbelievable. Some days you wonder how the net
can hold so many fish. But there are no crabs. You figure it out."

"Yeah, we haven't had no stingin' nettles this year. I've only seen a couple of rays, and the huge rock aren't around. Too much freshwater is keeping them all down south," Charlie says. "We've had stingin' nettles so thick you could hardly pull up the net."

I ask about Andy's going back to school and Leroy's age, expressing the concern that the marine industry, like the skilled construction trades, is facing a severe shortage of skilled young craftsmen.

"It's a serious problem," Tommy says. "Young people don't want to work, or at least with their hands. We have a hell of a time in this business. I think we'd be up the creek without the Mexicans and Vietnamese. Honest to God, they are saving us. They work hard, providing the labor we need, and they are buying the cheaper food fish—croakers and the like."

As Andy comes into the cabin Charlie adds, "Andy never worked so hard in his life as he did today. You're brown-nosing your way into Mick's book ain't ya?" We all, even Andy, smile.

Fish are culled, or sorted, by type and size and put into bushel baskets aboard *Miss Cindy.*

"Damn right, I think I might just sign Mick on as crew, Andy, and see how hard you can really work. You didn't even fall asleep goin' up the Bay this mornin'!"

"Oh, I slept a little. You can believe that," Andy said sheepishly.

Leroy stuck his head in the door and yelled, "Yeeoww," for no reason, except that later he would explain how happy he was that he would get to see his soaps. "I love *The Young and the Restless*. But I'll watch 'em all. They got some slimy crooks on those shows. I wouldn't miss 'em though. That one guy on *The Young and the Restless* is a real SOB." He then lightly punches Andy on the shoulder and gives him a jab, "You did pretty good today. Almost worked as hard as a girl," and the laughter starts all over again.

Andy immediately responded, with "Old man, I'm sweatin', aren't I?"

We all smiled, even Andy and Leroy, as we listened to the dynamics between young and old, building to another fit of laughter.

On this day, the crew culls perch, rockfish, sea trout, croaker, flounder, and an occasional bluefish.

The crew of *Miss Cindy* must reset the pound net after emptying the fish for the day.

Back at the dock, the baskets were unloaded, iced down, and loaded in Tommy's pickup truck to be taken immediately to the buyer in Jessup, Maryland. The boats were washed down and gear put away.

They unloaded the two fifty-five-gallon drums of bottom paint by the net dipping trough and started to say their good-byes for the day. Tommy asked Leroy and Andy if they were going to work tomorrow on gear, and they said they would.

Leroy asked, "What time, Cap?"

"Six o'clock," Tommy responded.

"Okay, let's go, old man," Andy said. With the work done, the young wanted to leave and the old had to follow.

Tommy asked me if I wanted fish to take home. I declined the offer with great appreciation telling him that on a poor day I wasn't going to dip into the pot. He insisted, and I insisted that I wouldn't. We settled on a shirt that said "Work is for people who don't know how to fish," and the promise that I would be back out with them in September or early

Charlie Quade puts away the large dip net used to remove fish from the pound net.

October when the fish might be more concentrated and not so scattered around the Bay.

Actually, the fishing did improve. Tommy called the following Monday to report that he took four thousand pounds of croaker from one net. Captain Russell Dize, now off his skipjack and working RDS Seafood, reported the same success with pound netters on the Eastern Shore.

On Labor Day weekend I called Tommy to ask if I could bring the former CFO of the National Marine Manufacturers Association, Bob Harris, out to see the netting operation. Harris is from Chicago and had never been on a commercial workboat. Hallock said, "Sure," and told me to meet him and the crew at five thirty Friday morning. I have always found watermen to be gracious hosts on their boats, welcoming strangers to their way of life without reservations. This is their opportunity to discuss their issues and concerns and allow their guests to see how they earn their keep.

While it was not a very successful day, Bob Harris returned to Chicago with a deep appreciation for Tommy, Leroy, and Andy. He enjoyed their company, the spectacular fish, and the information Tommy readily relayed to him. Driving back from the trip, Harris mentioned

Charlie Quade gave up crabbing to fish with Tommy Hallock. Here, he tags a rockfish, a requirement that must be met before leaving the boat.

how pleasantly surprised he was at the generosity and kindness of the watermen. He brought up the three fish Tommy gave me to give to Cindy's parents for her father's seventy-third birthday. Harris was also surprised that Tommy gave fish to Bob Evans to supplement his lost income due to an injured ankle. Knowing the nature of other watermen and how much Harris enjoyed this trip, I told him there were other men I'd like him to meet on another trip to the area.

CHAPTER 14

Captain Larry Simns

For over a quarter of a century Larry Simns, sixty-three, a fourth-generation waterman from Rock Hall, Maryland, has served as president of the Maryland Watermen's Association. He usually serves by default because no one else is willing to take the job. One of the conditions of being president of the association is that the individual has to be a full-time waterman. This makes it difficult because there are few commercial watermen who can fulfill the job responsibilities of an association executive and even fewer who would try. Simns, who has a bad back and becomes increasingly weary from the commute to the Annapolis headquarters of the MWA, continues to endure and has managed to survive. He remains a successful charter captain and gillnetter to fulfill his commitment to the MWA bylaws. And, according to many in the society of watermen and outsiders from legislators to the media, Simns has been a constant force ensuring that commercial watermen are able to retain their way of life.

I have known Larry Simns personally and professionally for over twenty years. I have witnessed him in state and national fishery-related legislative and regulatory arenas. Many people, like me, learn from him by observation, hoping to absorb some of his wisdom, political negotiating skill, and wit so that we might be half as successful or influential with our own efforts.

At about five feet, eight inches tall, slight of build, and wearing glasses, Simns is, by all accounts, nonthreatening and unassuming. He speaks in a soft Eastern Shore waterman's dialect, and he weighs his words regardless of his company. His conversations are frequently about what he does: promoting and protecting the Maryland watermen and their way of life.

He is calm in the face of most political or regulatory storms, using his talent for analyzing the situation and demonstrating his art of negotiation which leaves his opponent confused as to where the argument is going. He usually directs the conversation to right where he wants it. After the person across the table *thinks* he has either won or struck an adequate deal, it's then that he realizes that, while it might be true for him, Simns definitely got what he wanted. He always seems to know what his opponents are thinking and where they are going before they reveal the total picture. He can then steer them into his train of thought. He is not malicious in his manipulation; he's only looking out for what he represents.

A good example of Simns's technique is revealed in the debate over the dumping of dredge spoil in the deep trough off Kent Island and just south of Site 104. While the watermen have always been against open Bay dumping, and particularly in the deep trough where they do winter fishing, Simns saw the political opposition to the watermen, so he negotiated with state leaders. He suggested that if the commercial fishermen located another suitable spoils site, the government should immediately begin to give them $10 million for the oyster replenishment program and agree not to dump spoil in the deep trough. The state leaders needed time to discuss his proposal, but they finally agreed.

From the beginning Simns had a location in mind, Site 104, which already had been used but had room for more dredge spoil. "This would have been good because the original stuff was all toxic from the Baltimore Harbor and the new material coming from the channel could cover it up. I knew this was a good alternative going in. I just didn't play

Larry Simns, president of the MWA, is surrounded by watermen as he prepares to testify before a committee on a body of legislation. This is Simns's full-time job between January and May. Courtesy Maryland Watermen's Association.

all my cards until I had the agreement. See, state people change so often they don't know the history. I kept this in my back pocket. At the end of the day, Site 104 was shot down and we still got our money [$3 million] for oyster replenishment. We won all the way around," he said.

In the early nineties when the state had completed a five-year moratorium on rockfishing, the Maryland Saltwater Sportfishermen's Association (MSSA), hired renowned state lobbyist Bruce Bereano to represent them and to get the game fish status for the striped bass. On the day of the hearing, there were so many people wanting a glimpse of this encounter that the chairman had to move the meeting from the normal committee room to the state legislature's large hearing hall. Several hundred attendees, many having come directly from their boats to make the start of the meeting, sat theater-style and waited in anticipation to see how Simns would face Bereano. A variety of people attended: roughly equal numbers of watermen, sportsfishermen, and rubber-neckers; professional lobbyists, General Assembly staff, Natural Resources bureaucrats, the media, and more.

Dressed in a navy blue suit, Bereano shook hands with members of the committee and with people in the audience. As he worked the crowd, he looked every bit like the highest paid lobbyist of the Maryland corps, making over a million dollars a year. Bereano testified first for the proponents.

Simns, in a blue blazer, looked at ease and every bit like a waterman out of his natural element; he testified second, for the opponents. The few witnesses were inconsequential; attendees and legislators came to hear only two individuals. At the end of the day, Simns defeated the proposal, beating the MSSA with such finesse that they never realized it had happened. Many will remember how Larry Simns handled himself and represented the interests of the watermen. I have seen him in action many times since, and it is always exciting, whether on the state or congressional level.

In June 1999 Simns took his fears and concerns from the past decade and for those ahead to his colleagues. In an open message to members of the MWA he wrote a column in the *Watermen's Gazette* entitled: "Are watermen under siege? Simns takes a hard look at the opposition." Here are some cogent extracts from that column that put his thoughts in perspective:

"Lately it seems as though the Watermen's Association finds itself in far more battles than it ever has in the past. . . . It appears to me that every activist who forms an organization wants to do it on the backs of watermen. These groups, too many to count, disguise themselves as being for the environment, or some other issue, and not necessarily against watermen. Experience has shown us that the ultimate outcome is yet another slam on our livelihood.

"The waterman is our society's last free spirit, last free hunter, a fact that this same society can't seem to accept or respect. What is not easily understood is condemned by those who do not have the ability to see our way of life as a good way of life. Our opponents go so far as to blame us for the detriment of the Chesapeake Bay, when in actuality, watermen have been a primary force in keeping the Bay in relatively good shape. Whether it's the realtors who sell to developments, sewage treatment plants that contaminate our waters, or the sports fishermen who pollute with their boats, we are all to blame!

"And, these factors are just the tip of the iceberg. The list goes on and on: dredging of the channels to dump mud, industrial polluters, and farm run-off . . . whether agricultural or livestock . . . from large factory farms. Big business wants to eliminate us because of the very nature for which they stand.

"Presently, there are groups whose main initiative is to eliminate gill nets, the tools watermen use during winter months to make a living, all the while claiming they don't want to hurt us. Their representatives completely ignore the fact that the MWA has gone to great lengths to place restrictions so that we don't overharvest, so that we stay within the guidelines of an allowable catch, and so we don't waste anything.

"First, they will do away with gill nets. Next, it will be pound nets. Soon after, it will be commercial hook and line. Eventually, as these groups need a cause to stay in existence they will want to do away with fishing, holding fish, or even disturbing fish!

"These groups send out propaganda to newspapers to gain membership, as do some environmental leaders who create crises to boost membership and donations, and this is done on the backs of watermen."

He says that groups who want to save the submerged aquatic vegetation, in their zeal, are on a road to put commercial clamming out of business. Next, Simns says, it will be the crab scraper. "To me, it is wrong-minded to take something away from another group merely to justify one's own existence. Furthermore, most of these activists have already made a good living and don't have to worry about someone fighting to take their livelihood away from them. They don't know what it means to be down on our level, working day to day to survive in what we do."

Simns explains that real estate brokers who sell prime farmland and prime waterfront property to developers are perhaps doing the most damage to the Bay and to the watermen. It puts a strain on the resources, and the people who buy the expensive property want to control some of the poorest people in the community, taking away their way of life because it interferes with the new residents' pleasures. "They have the time, money, and energy to devote to catching the last fish out here with a hook and line, but don't want us to make a living off of these same fish!

"Look at it this way. These recreational fishermen talk at hearings about how detrimental commercial fishing is because watermen are catching fish and selling them. Well, a dead fish is a dead fish, whether you catch it and sell it or catch it and eat it, or you catch and release—and ultimately kill it—at least 8 percent of what is hooked dies. If you took 8 percent of the hundreds of thousands of recreational fishermen and hundreds of thousands of fish they catch, that amount exceeds the total catch of the commercial fishermen in the Bay. Then again, these new waterfront property owners hate the clammers because of the noise they create offshore!

"The bottom line is always the same," Simns says. "Watermen need to know what is happening to our livelihood, and the people who are attacking us need to read this and realize the effects of their actions. There are some straight shooters and legislators who are very fair-minded, but these are the people the other groups are trying to rid from the political scene. The MWA will continue to fight to ensure our way of life is protected so that Maryland does not go the way of Florida, Louisiana, or any other state in which fishing rights have been taken away from watermen. We hope you will join us in this fight."

A year later, I caught up with Larry Simns. It is easy for me to reach him on his boat or at home, but nailing him down for an extended period of time is nearly impossible. We agreed to meet in his hometown of Rock Hall at the Pasta Plus restaurant at noon on October 4, 2000.

Driving across the Chesapeake Bay Bridge and north on Route 301, I notice that the new housing developments in Kent Island and Queenstown quickly give way to large farms with corn standing as brown sentries stretching as far as the horizon. Once off Route 301 and on Route 213, I observe a landscape that remains the same with only two major breaks, the towns of Centreville and Chestertown. Rock Hall sits at the end of the road, so to speak. It is surrounded by farms and water, and except for the overhaul of its waterfront to accommodate condominiums and marinas, it has changed little in the past fifty years. It is a waterman's town and regardless of where you go on any given day you will see men from the water.

When I walk into Pasta Plus, Simns is sitting at a table for six in the far corner. There are three other watermen sitting with him and several seated at surrounding tables. He waves me over and I am offered an empty chair beside him. I know, by face, several of the men at the table and after exchanging greetings with them I am introduced to the others. Most of them have been taking charter-fishing parties out because crabs and clams are in such bad shape.

As Larry and the others finish their meals, I order and follow the discussions with interest. Their conversation moves from one subject to another. They speak of Al Gore: "We vote with our trigger finger [they are opposed to gun control], and can't support him 'cause now he's gone too far with his liberal politics." They also speak of a local waterman whose wife drives a BMW but won't let him buy a new engine for his boat: "When a woman starts telling you how to run your business it's time to leave her. Two things go wrong when you change your ways because of a woman. First, you ain't happy, and second, even if you change your mind or your ways, she ain't goin' to be happy. It's best to cut it off. Hell, at one time I had an ex-wife on every corner of this town, and we're all still good friends," Simon says with a smile.

As men get up and leave, others join the group. Larry seems to be holding court and the variety of issues continues for another hour until he gets up from the table. He waves me on and I meet him outside after paying my bill.

"Get what you need. You can ride with me," he says as he walks toward his pickup truck. A three-year-old golden retriever is in the back and goes crazy with excitement when I approach the truck. "That's Cap.

Pet him or he'll drive you nuts until you do. He loves people," he says as we both pet Cap before getting in the truck. I have no idea where we are going or what we are going to do. He agreed to spend time with me, but I am on his turf now and I know he has a plan. He always does.

"We'll go over to the beach at the public landing. I want to throw a toy to Cap and wear him out. We can talk then."

Following are some of his thoughts on issues facing the watermen of the Bay:

ACCESS

"Access is a major issue for us because it's a problem. Nobody wants us around. We have more slips and public landings that watermen can use than probably any county in the state and we're planning for more. We've pretty much stopped all condominiums from encroaching on us. See, marinas are fine because we can work there when things get slow but condominiums don't do anything for us except take up the waterfront.

"Now, this place here is called Eastern Neck Island National Wildlife Refuge. It's state land now but when I was a boy I used to work for hunting parties out here in the winter. I've combed every inch of this SOB in daylight and the dark. I've hunted on it and worked on it. Now, years back we had a state senator who managed to get this purchased by the government. We were all pissed off because it meant we lost our jobs. It's the best thing that ever happened because if the government didn't buy it up we'd have condominiums comin' out our ass."

DEVELOPMENT

"Development with all its runoff is what's hurting the Bay. The Bay's so enriched with nutrients that we have the algae blooms and fish kills. That's bad but people don't get upset over what they can't see. Why, sewage treatment plants are the worst things we have on the Bay. First, they allow for places like Rock Hall or Tilghman Island to be overdeveloped which puts a strain on the natural scheme of things as well as our local people 'cause they can't afford to live there any

Captain Larry Simns with his dog Cap and a friendly admirer.

more, but look at the sewage spills. Millions of gallons goin' into the rivers and into the Bay and no one does anything about it. If we pump our bilge and have a trace of a slick on the water and someone blows the whistle, the Coast Guard will give us a $10,000 fine and surround our boat with pollution control equipment like a circle of covered wagons. Sewage treatment plants haven't solved problems; they've created them, and the laws are made for little people, not the big corporations or the cities."

ELECTIONS AND POLITICS

"I try to get along with all the legislators—local, state, and federal. You can't take any of 'em for granted and you've got to work 'em all the time. The absolute biggest challenge we have is to keep government regulation from shutting us down. The other is to keep the Bay clean so we can have successful reproduction of the species. This means we've got to stay up on the elected officials and be able to siphon out the good science from all the junk science out there that's used against us by officials."

RECREATIONAL INDUSTRY

"You know better than anybody that I've always tried to defend our position or interests without fightin' them. What I can't get people to understand in my industry or the recreational industry is that we've got to work together or ain't neither of us goin' to win; either the government or the environmentalists will. Actually, we've always gotten along right well with the recreational industry. Boatyards have saved us with jobs when men had to get off the water for a time. I suppose if I have a concern it's that recreational anglers think we take the fish and overharvest when in reality they probably take as much or more than we do. And it ain't overharvesting. The resources always move in cycles and what's impacting them now more than anything is environmental, not commercial watermen. It can't get down to a race for the last fish between commercial boys and recreational fishermen. We've got to put our heads together to preserve the resources so we both have a future."

THE MARYLAND WATERMEN'S ASSOCIATION

"I would have been gone some time ago if I had somebody to take my place. I've got different people in training now and it might take more than one person to do the job. Things are different nowadays. What's for certain is that the Watermen's Association has got to continue and be even stronger. It's the only thing between surviving and dying in the watermen's way of life. Boys back off because of the time and energy it takes to be president and the money ain't real good. I think we have good credibility and a good staff right now but it's hard keeping good people because we can't pay much, but we can't afford to let her die either."

REGULATIONS

"As I said, regulations are what will shut us out of business. See, this is a young man's job. You have to have the strength and willpower to hold on and work long, tough hours. But regulations restrict a man's time. Watermen will work from sunup to sundown to try and make it and if you show them the money they'll find a way to get at it. But regulations take his ability to work hard away from him, and they shouldn't. That's why we've got to keep workin' the state and the legislature: to give men an opportunity to survive."

WATERMEN

"We're a creative bunch and most of us aren't suited to do anything else. We don't want to do anything else. We can work full time at what most people enjoy as a hobby—working on boats and engines and be out on the Bay. We can figure out how to get by doin' somethin' if the government doesn't take our ability to work away from us. We've got a waiting list of four hundred for the apprenticeship program and over a thousand to get a license. So it's not a dying culture. Now, some boys will get off the water temporarily but the truth is most of 'em don't stay off it. They don't want to or they can't. If you owe between $100,000 and $200,000 on your boat and equipment you've got to have that boat workin' to pay for its share. Besides, in bad times nobody's goin' to buy the boat. Sure, there were men that got out of crabbin' this summer and put their pots on the shore, but they're on the shore. They ain't sold 'em. They'll be back crabbin' next year and it might be a lot better. It's always fluctuating. Ain't a person on this earth that can figure out what a crab is goin' to do so we're always sidin' with optimism.

"There's all kinds of watermen, too. Some go for where the money is and move from prosperous harvest to prosperous harvest. Then if it dries up they move off the water. For most, it's hustle and hard work for the young. It's always been that way. About half make it and half don't. But you've got to remember that out here if you gross $100,000 in a year you're only bringin' home about a quarter of that. The rest goes to crew, the boat, and equipment. So if a businessman were to look at what we do he'd think we were crazy. Maybe we are. Who knows? We do know that we love what we do. Young men have to learn that, in the good years, you've got to put money into your boat and equipment so that they can coast with their boat in the leaner times. Now, for example, is a lean time and those that make it are coasting because their boat and equipment are in good shape. They'll be ready for when it picks up. If you don't take care of your boat and equipment, your business, you're done. It takes plannin' and that's what we try to do."

As he takes me on a tour of one public landing after another, tossing Cap's toy into the water at each one, we near the time when we have to head in different directions. I'm going south toward home, and he's going target shooting with his preacher. I ask him what he thinks about the future for the watermen. "I'll tell you that I'm optimistic if we can stave off regulations and work with the Chesapeake Bay Foundation and others so species can reproduce, then we'll prosper. People want to eat seafood. They need us. So as long as we can maintain our ability to work we'll be okay. That's what the association is for. Right now, I'm worried about clams—they have Dermo, but a different strain than the oysters—and crabs. I'm worried about the loss of grasses and increased pollution. And I worry that we get blamed for things we don't do because we're an easy target. This is an uphill job and way of life but I'm optimistic. It's all we know."

On our way back to my truck at the Pasta Plus he pulls off the road to another public dock where the old oyster buyboat *Rebecca Forbush* is tied up. She is owned by Captain Charlie Marsh from Smith Island, and she looks beaten up although the oyster season just started. Charlie follows the fleet around and buys oysters from the men to ship out of state. The men can harvest fifteen bushels per license and have two licensed men on board for a maximum of thirty bushels. Charlie pays $21 a bushel today, and as boats come in, they unload their catch by the bushel. From the boat they are put on a conveyor belt set up amidships of *Rebecca Forbush* and carried into one of his trucks. These oysters will go to Virginia. Sometimes the oysters are transported directly from the workboats to the buyboat's deck as they were in decades past. Now there are only a handful of the big boats left and Charlie is pleased that Larry and I stopped by to see his boat at work.

"This is one of the few buyboats left. It's a shame. If the government had put money in the oyster replenishment program years ago we'd have oysters as big as basketballs and plenty of 'em. Now we've got to run all over the Bay findin' oysters. This could be a good year. They look good comin' out of the water this early in the season so we'll all keep our fingers crossed," he says as he keeps track of the oysters being unloaded by marking the bushels down on a legal pad next to the captain's name. I can see that at least six boats have been by already and his truck is only half full. I hope he gets a full load as we walk off his boat and head back to the truck. Oystermen pay $400 for their license plus a $300 surtax that goes to the oyster recovery program. Charlie and other buyers pay

Larry Simns at a public dock in Rock Hall, Maryland. Through his efforts, the town has provided slips for commercial workboats away from recreational yachts.

Workboats at rest at a public dock in Rock Hall, Maryland.

$1 a bushel surtax into the program. It still costs money to make money and to keep the resources around for the next generation.

After spending the afternoon with Larry Simns I leave him in the restaurant parking lot, go in and get a cup of coffee for the road, and plan to spend the long drive back through the country contemplating Larry's hopes and objectives for the watermen of the Chesapeake. Like it or not, he is the nerve center of this small society and much revolves around and relies on him—perhaps more than he himself realizes.

Epilogue

In April 2001, Jim Price wrote a commentary in the *Bay Journal* declaring that reports from the year 2000 could leave no doubt about the health of the Bay—the Chesapeake was an ecological disaster. Water quality was so poor in some major tributaries that record-low dissolved-oxygen levels were recorded during the summer, after one of the largest and most damaging mahogany tides ever observed. Numerous fish kills occurred throughout the Bay region. As much as 50 percent of the striped bass population was affected by a slow-growing, progressive disease which caused an infection that attacked the internal organs of the fish. In addition, the crab harvest of 2000 was the lowest in recorded history, forcing many watermen to quit early, and even forcing some out of business entirely. Without a healthy and productive estuary, watermen *have* no business.

In addition, the Chesapeake Bay Foundation released a statement saying that $8.5 billion would need to be spent over the next ten years, or $850 million annually, to create programs that would curtail pollution and meet the goals established in the tri-state agreement signed in 2000. The money would be used to preserve thousands of acres of watershed property, to create buffer zones to protect rivers and streams, and to reduce pollution from sewage and urban storm water. Funding would come from all levels of government and from nonprofit organizations.

Few people would question the importance of maintaining the Bay's health. The Chesapeake is the largest estuary in the United States and touches six states. It is considered to be among the most diverse in the world, with more than 300,000 species of aquatic plants and animals. However, the new Bush administration *is* questioning the amount of money it can budget to meet what CBF claims the Bay requires. The president has said that he will propose trimming the Bay's funding, but the proposal will not even come under congressional scrutiny for months. Despite the president's position, Bay proponents usually get the money needed to improve the quality of the Chesapeake, so there is still hope that the watershed will receive full funding.

While the health of the Bay is critical to the livelihood of the watermen, it is equally important that the watermen not lose the freedom to pursue that livelihood. When the Maryland legislature ended its session in April 2001, the Maryland Watermen's Association released a report noting that the group's lobbying efforts had affected twenty-one bills dealing specifically with waterman issues. Ten bills passed and were sent to the governor to be signed into law. Of these, two require recreational crabbers to obtain a crabbing license, one prohibits the importation of nonnative crab species into the state, and one requires a mandatory day off per week for commercial crabbers.

None of these laws was momentous from the association's perspective, but what was significant about the 2001 legislative session was the number of bills that were withdrawn by their sponsors or that failed to pass out of committee. The demise of these bills shows the influence of the watermen's lobby. Consider the bill that would have put a catch limit on crabs after October 1, the bill that would have restricted days and areas for crab harvesting, or the bill that would have raised the size of a peeler crab (soft shell) from 3" to 3½". Another proposed bill would have given the Department of Natural Resources search-and-seizure power in connection with a suspected offense, made such offenses criminal charges, and given the department broad authority to enforce laws

In the harbor of Rock Hall, Maryland, a sculpture of a hand tonger pays tribute to the men who work the water amid crab pots and oyster boats.

that govern the commercial fishing industry. All of these bills would have significantly impacted the lives of the watermen, but the lobbying by the MWA successfully prevented them from passing.

That's good new in the short term but watermen know their life-style—their subculture—must change with the changing environment. These native hunter-gatherers watch fluctuations in the resources, as they have for centuries, but now it is different. Not long ago they could count on preserving their livelihood by moving to a more abundant fishery when their primary target let them down. Today, there are no guarantees, and too many men and women are leaving the water to secure their financial futures.

While there may always be a commercial fishing industry on the Chesapeake, it may not be identified by the same symbols it once was: small towns of working watermen, shucking and packinghouses built on oyster shells, and children running to meet the boats when they return to the dock. Many of those towns are now occupied by out-of-towners; packinghouses are all but gone; recreational boats fill slips; and children are encouraged, perhaps reluctantly, to find vocations off the water.

The industry may not be identified by thousands of pickup trucks moving across the state with bushel baskets or refrigeration cabs in the back and bumper stickers saying "I'm proud to be a waterman" or "Don't be a crab or a waterman 'll get ya," or even by men wearing white rubber boots. Those trucks and men will decline in numbers leaving only those who have known the water for many years and who planned for the future by saving enough money to survive during the slow years. There will be fewer young men and women who speak a unique dialect and who remember learning the water from fathers and grandfathers and from oral histories woven deep in the small-town fabric of family. It is too expensive to buy a boat and gear, regulations are too numerous, and opportunities on land too diversified and compelling for young folks to ignore. They've seen what has happened to too many fathers, brothers, uncles, cousins, and neighbors.

For now, watermen still exist, and they will be able to tell us about the changing Bay. They will try to tell us what is wrong with the estuary and how to correct it. Unfortunately, my guess is that they will be educating us through their fights to divert blame from themselves for overharvesting to the real problem: all people. The Bay is continuing to change because we allow it. For some, the change is viewed as an improvement. Others see it in a state of gradual decline. Regardless of the perception, the continuing influx of people in close proximity to the Bay and her tributaries and the loss of grasses will cause extraordinary pollution and will have a strong impact on natural and human resources.

Perhaps this book illustrates that watermen may be an endangered species. Should their culture become extinct, much like that of the

When all is said and done,
he'll hang it up and be back
tomorrow to do it again.

Native American, scholars will have to do research and thesis work by
relying on oral history and a few remaining watermen. And, in Bay
country, we may see a growing subculture of people driving pickup
trucks and wearing white boots to try to identify with or to reclaim
what we've lost. They may even be the children of the subculture so
prevalent years ago, who tried in vain to identify with and save Native
American tradition and spirit. It would be wiser for us to learn from
history and from our mistakes and try to preserve the watermen and
their way of life now. This can be achieved by controlling the resources
and allowing the watermen to survive in their environment rather than

forcing them to transform and then trying desperately to reclaim their way of life when it's too late.

I for one will continue to try to understand the watermen, to absorb all the knowledge and wisdom that I can from them, and to pray that they do not continue to fall victim to politics and greed. I know they are not the reason for declining harvests. It's just easier to blame them and make new assumptions, regulations, and laws rather than to see that all people are contributing to the condition of the Bay and must share responsibility for its decline or improvement.

The Bay will continue to undergo tremendous pressure from pollution and development because politicians will continue to be pressured by developers and big business. Fishery stocks will continue to fluctuate as they have for decades because of overharvesting, natural problems, accidents like sewage spills, and competition between recreational and commercial anglers. Watermen will hold on as best they can to the only world they know, despite the ebbs and flows of a changing ecosystem, political pressure from sophisticated adversaries, and the tremendous anxiety over the eventual retirement of Larry Simns and Betty Duty from the watermen's one security blanket, the Maryland Watermen's Association.

As I continue to study the watermen, I will think of the State of the Bay report that said the watershed has not made significant improvement since the last one in 1998, except for an increase in shad. But I will also think of Jimmy Iman, Jr., from Fort Howard, Maryland, who has worn the waterman's uniform of white boots, jeans, and a baseball cap for over eight years. He's been a licensed waterman since that time and he can gauge salinity, navigate by the stars, lay a 450-foot trotline and break down an outboard motor with the best of them. Jimmy speaks freely about the difference between a peeler and a soft shell. He has mastered many of the techniques that distinguish old-timers from mere imitators. Jimmy Iman loves the water and the life of a waterman.

Jimmy is thirteen years old. He deserves his opportunity to make his mark. A decade from now I hope he will still be out there in his boat, at age twenty-three, saying, "There's nothing I like better than being on the water."

Index

Abner, Bobby, 201
Anne Arundel County Watermen's Association, 31, 32, 201
Arundel Marine Construction, 167
Ashley, Danny, 215, 219, 220
Atlantic States Marine Fisheries Commission, 11, 69–70, 167
Austin, Herb, 68

Baker, William, 200
Baltimore County Watermen's Association, 121, 127
Beck, Danny, 51, 99, 121–41, 181, 205, 221, 227
Beck, Joyce, 99, 121, 122, 124, 129
Bereano, Bruce, 121, 245–6
Bevans Oyster Company, 119–20
Bi-State Blue Crab Advisory Committee, 20, 80, 183, 187, 195
Blackistone, Pinkney, xi
Bowers, Harvey, 214–22, 223, 225
Brown, Robbie, 106, 110–9
Brown, Robert T., 106, 108
Brown, Torrey, 129–30
Brumbaugh, Robert, 101, 184

Caldwell, Rita, x
Calvert County Watermen's Association, 14, 201, 205
Chesapeake Appreciation Days, 3–4, 7–8
Chesapeake Bay Commission, 163, 183, 195, 197
Chesapeake Bay Foundation, xii, 31, 36, 45, 69, 74–81, 101, 110, 184, 187, 197–200, 213, 254, 257, 258, 262
Chesapeake Bay Institute, 201–2
Chesapeake Bay Program, 9, 164, 194, 210
Chesapeake Bay Seafood Industries Association, 56
Chesapeake 2000 Agreement, 79–80, 164, 194–7
Chones, Terry, 84
Coastal Conservation Association, 31, 88, 131, 139
Crewe, Linda, 98–102
Crewe, Ronnie, 100
Crook, Bill, 222–5
Cummins, Jim, 72

Cummings, Bill, 5
Cummings, Jean, 5

D'Amato, Richard, 55
Department of Natural Resources. *See*
    Maryland Department of Natural
    Resources
Dize, Leah, 146
Dize, Russell, 8, 46, 142–59, 182–3,
    240
Dize, Rusty, 8, 145–6
Donoffrio, Jimmy, 11
Drake, Susan, 75–7
Dunn, Jaimie, 99
Duty, Betty, xii, 10–2, 19, 34, 83, 88,
    90, 101, 188, 260, 262
DuVall, Dave, 225

East Coast Commercial Fishermen's
    and Aquaculture Exposition,
    82–91
Eastern Neck Island National Wildlife
    Refuge, 250
Echo Hill Outdoor School, 45
Elburn, Danny, 49
Emmert, James, ii, 30, 33, 34, 35, 36,
    37, 38, 41, 43, 44, 51–2, 57–64
Evans, Bob, 13, 29–45, 51–4, 57–64,
    110, 121, 131, 187–8, 201–5, 228

FAIITH (Families Actively Involved
    in Improving Tangier's Heritage),
    76, 81
Fithian, Ronnie, 93–8
Fitzgerald, Dan, 122, 123, 124–5,
    132–5, 138–9
Flood, John, 40

Gilchrist, Wayne, 191
Glendening, Parris, 9, 191
Goldsborough, Bill, 69, 79–81, 187
Gross, J. R., 188

Gutting, Richard, 20

Hallock, Tommy, vi, 227–9, 233–42
Harris, Bob, 240–2
Harrison, Bob, 240–2
Harrison, Captain Buddy, 5, 47–8, 91
Harrison, Captain Buddy, Jr., 5, 91–3
Harrison, Leslie, 5, 91
Harrison, Verna, 129
Hastings, George, 10
Hewison, Adam, 201–5, 228
Hickman, Jeff, 182
Higgins, Eddie, 215, 222
Hirshfield, Mike, 213
Hughes, Jane Blackistone, 56

Iman, Jimmy, Jr., 260–1
Interstate Commission on the Potomac
    River Basin, 72

Jenkins, Doug, Sr., 185
Jensen, Pete, 45, 73, 131
Johns Hopkins University, 32, 201,
    228
Jones, Amos, 31, 34, 35, 36, 38, 39, 41,
    42, 43, 51–2, 57–64
Jones, Phil, 70
Junkin, Tim, 85

Keen, Karen, 12, 15–7, 85
Keen, Kenny, 14–29, 38, 40, 43, 85,
    128, 140, 183, 201, 205
Keith Underwood and Associates, 167
Kelly, Linda, 201
Kittleman, Robert, 191

Lassahn, Karen, 99
Lassahn, John, 99
Legg, Martin, 17
Leggett, Vince, 42
Leibowitz, Elissa, 207–8
Leik, Elizabeth Scott, 222–3

Lingerman, Chris, 49–50
Lipcius, Rom, 184
Lipton, Doug, 186
Long, Joe, xi

Macon, Clementine, 10
Maddox, Walter Irving, Jr., 103–19
Maddox, Walter Irving, Sr., 13,
    103–19
Marina Operators Association of
    America (MOAA), 20, 32
Marine Recreational Fisheries Statis-
    tics Survey (MRFSS), 206
Marine Trades Association of Mary-
    land (MTAM), 20, 32, 95
Marsh, Chuck, 81
Marshall, Janice, 81
Maryland Association of Counties, 88
Maryland Department of Agriculture,
    176
Maryland Department of Economic
    Development, 48
Maryland Department of Natural Re-
    sources, 8, 25, 33–4, 40, 45, 49,
    66, 70, 71, 73, 78, 88, 97, 112,
    126–7, 129–33, 162, 165, 201,
    209, 258, 262
Maryland Oyster Roundtable, 9, 20, 25
Maryland Port Administration, 48,
    193
Maryland Saltwater Sportfishermen's
    Association (MSSA), 245–6
Maryland Seafood Marketing Advisory
    Commission, 35
Maryland Tidal Fish Advisory Com-
    mittee, 48
Maryland Watermen's Association, xii,
    6, 8, 10–2, 15, 16, 19–20, 29, 34,
    82, 83, 85, 94, 120, 121, 129, 140,
    162, 188, 191–3, 218, 221, 243,
    252, 258, 262
Matuszeski, Bill, 210

McLaughlin, Shawn, 213
McNasby, Mac, 15
Minkkinen, Steve, 71
Morgan, "Tootsie," 106, 108, 111–9
Murphy, Bart, 151
Murphy, Lawrence, 151
Murphy, Wade, 46–8, 92

National Blue Crab Industry Associa-
    tion, 170
National Fisheries Institute (NFI), 20
National Marine Fisheries Service, 20,
    165, 210
National Marine Manufacturers Asso-
    ciation (NMMA), 20, 240
National Oceanic and Atmospheric
    Administration (NOAA), 70, 213
Neall, Robert, 191
North Carolina Fisheries Association,
    165

O'Donnell, George, 86–90, 190, 193
Orme, John, 32, 33, 40, 201
Orner, Derek, 70
Orth, Bob, 216
Owens, Janet, 204
Owings, George, 186
Oyster Festival, 10
Oyster Recovery Partnership, 8–9, 20,
    84, 142

Patuxent River Commission, 20
Pfeiffer, Bob, 8
Philip, Leroy, 229, 230–2, 234, 238, 239
Phillips Seafood, 54–6, 131–2, 188
Phillips, Steve, 132
Poukish, Kent, 212–3
Price, Jim, 257
Pruitt, Ken, 77, 101

Quade, Charlie, 44–5, 140, 183, 228,
    230, 240

Queen Anne's County Watermen's Association, 162

Recreational Fishing Alliance, 11
Rice, Billy, 99
Rice, Tina, 99
Rock Hall, Md., 94–5
Rock Hall Watermen's Festival, 3
Rohlfing, Joe, 122, 123, 124, 125, 129, 132–9
Russell, Jackie, 13, 45–6

St. Clement's Island Maritime Museum, xi
St. Michaels, Md., 4
St. Pierre, Richard, 210
Save Our Skipjacks Task Force, 48
Schill, Jerry, 165
Seiling, Bill, 176
Sheckells, John, 34
Sherbo, Billy, 176
Shores, Carlene McMann, 75, 76, 78
Shores, Rudy, 78
Simns, Larry, xi, 10–2, 19, 25, 33, 34, 56, 69, 81, 88, 90, 94, 101, 120–1, 127–8, 140, 163–4, 191–2, 205, 218, 221, 243–56, 260, 262
Slaff, Bob, 85–6, 95
Smith, Barbara Beck, 171
Smith, Scottie, 17, 171, 180
Smith, Steve, 17, 171–81
Sneed, Mark, 55–6
Spangler, Terry, 99
Stilwagen, Richard, 185
Strittmatter, Faye, 167
Surrick, John, 165
Swanson, Ann Pesiri, 194, 214

Taylor-Rogers, Sarah, 49
Thompson, Jack, 183
Tilghman Island, Md., 3–5
Townsend, Karen Sue, 99

Twin River Watermen's Association, 185

U.S. Army Corps of Engineers, 9, 161–2, 193
U.S. Department of Agriculture Agricultural Research Service, 213
U.S. Department of Agriculture Natural Resources Inventory, 196
United States Environmental Protection Administration (EPA), 70, 164
University of Maryland Oyster Hatchery, 8–9
University of Maryland Biotechnology Institute, 186, 188

Vaughan, Doug, 210
Virginia Commercial Fishing Advisory Board, 101
Virginia Institute of Marine Sciences, 68, 184, 213, 216
Virginia Marine Resources Commission (VMRC), 76, 79, 100–2, 168, 183

Watts, Dave, 17
Watts, Kenny, 13, 17, 172–3, 176
William Clark and Son Seafood, 220
Willing, Jack, 223
Wilson, Brenda, 99
Wilson, C.R., x, 13, 15, 19, 98
Wilson, Jason, x, 13, 15, 19, 47, 98
Wilson, Robbie, x, 4, 5, 13, 15–7, 19, 98
Women of the Water, 99
Working Watermen's Association (WWA), 99
Wood, John, 183
Woodall, Ernie, 106, 108, 110, 111–9
Woodfield, Bob, 56

Zohar, Jonathon, 186